CAMBRIDGE STUDIES IN ECOLOGY

Editors:

R. S. K. Barnes *Department of Zoology, University of Cambridge*
H. J. B. Birks *Botanical Institute, University of Bergen*
E. F. Conr *Department of Environmental Science, University of Virginia*
J. L. Har *School of Plant Biology, University College of North Wales*
R. T. P *Department of Zoology, University of Washington, Seattle*

The ecology of the nitrogen cycle

JANET I. SPRENT
University of Dundee

The right of the
University of Cambridge
to print and sell
all manner of books
was granted by
Henry VIII in 1534.
The University has printed
and published continuously
since 1584.

CAMBRIDGE UNIVERSITY PRESS
Cambridge
New York   New Rochelle
Melbourne   Sydney

Published by the Press Syndicate of the University of Cambridge
The Pitt Building, Trumpington Street, Cambridge CB2 1RP
32 East 57th Street, New York, NY 10022, USA
10 Stamford Road, Oakleigh, Melbourne 3166, Australia

First published 1987

Printed in Great Britain at the University Press, Cambridge

British Library cataloguing in publication data

Sprent, Janet I.
The ecology of the nitrogen cycle. –
(Cambridge studies in ecology).
1. Nitrogen cycle
I. Title
574.19′214   QH344

Library of Congress cataloguing in publication data

Sprent, Janet I.
The ecology of the nitrogen cycle.
(Cambridge studies in ecology)
Includes index.
1. Nitrogen cycle.   2. Ecology.   I. Title.   II. Series.
QH344.S68   1988   574.5′222   87-13218

ISBN 0 521 32537 4 hard covers
ISBN 0 521 31052 0 paperback

**Transferred to
Digital Reprinting 1999**

**Printed in the
United States of America**

Contents

Preface

Since the nitrogen cycle was discovered it has been the subject of intensive research. In the last ten years the scale of this research has increased dramatically, largely in parallel with worries over nitrate pollution and acid rain. As a result of these efforts many meetings have been held and many symposia volumes produced, dealing with nitrogen cycling in different areas of the world. It is thus timely to attempt an overall summary of the current situation within the confines of a single volume.

In attempting this I have of necessity been very selective. Although in an ecology series, the book begins on a rather biochemical note, but a knowledge of the reactions involved is necessary before the effects of environment on these reactions can be assessed. Because so many different organisms are involved, it is impossible to consider each of them in detail. The emphasis is necessarily on microorganisms and plants, since the major reactions take place in them. However, in view of the increasing awareness of the role of animals (especially invertebrates), examples involving these have been included.

The second part of the book is devoted to case histories from different environments. The selection of examples may seem rather subjective, but the aim has been to cover as wide a geographical range as possible. This has meant the omission of much first-rate work from certain areas, such as North America. I apologize to all those who are not included and hope that references to relevant symposia will enable readers with specialist interests to find more examples for their particular areas.

Two aspects of nitrogen cycling have been largely excluded. First, the numerous models currently being drawn up have not been discussed as these alone could form a whole volume: one field example is included. Second, methodology is covered only in passing. It is one of the regrettable facts of current science that accurate field measurement of most nitrogen cycle reactions, but particularly nitrogen fixing and denitrification, is impossible because of limits in methodology. For these reasons amongst others, including the sheer volume of work involved, very few studies of nitrogen cycling include all descriptive (organisms, environments) and quantitative (rates of reaction, pool sizes) components. It has thus not been possible to adopt a uniform approach when discussing examples.

The book is intended for advanced undergraduates, early postgraduates and researchers in allied areas.

I should like to thank Professor John Harper FRS and Cambridge University Press for inviting me to write this volume. Discussions with many colleagues in various countries have assisted me greatly and I am particularly indebted to my Dundee colleagues Professor John A. Raven, Dr Linda Handley-Raven and Dr Rod Herbert. Members of my research group have shown great patience at my frequent absences when working at home. Without the help and encouragement of my husband Peter, who interrupted the writing of his own books to do all the word processing of the final version of the manuscript, the book would still be incomplete. I am indebted to the staff of Cambridge University Press for making the production of this book remarkably painless.

Janet I. Sprent

Dundee, Scotland
February 1987

PART I

General features of the nitrogen cycle

1

Introduction

Nitrogen is one of many elements involved in cyclical transformations in the world. However, after the carbon cycle, the nitrogen cycle is arguably the most important to living organisms. Organic compounds of nitrogen are probably as diverse as those of carbon, but, in addition, there are several inorganic forms of importance, compared with one – carbon dioxide or its hydrated form carbonic acid. Before considering the reactions of the nitrogen cycle, it is thus pertinent to consider what these compounds of nitrogen are and also where they are, in the major global compartments.

Forms of nitrogen in the biosphere
Atmosphere
Nitrogen gas
The dinitrogen molecule, N_2, makes up about 79% of the extant atmosphere, although this has not always been so. It is a very stable substance and a considerable quantity of energy is needed to break its inter-atomic bonds.

Nitrogen oxides
All possible molecules of nitrogen with oxygen may be found, i.e. N_2O (nitrous oxide), NO (nitric oxide) and NO_2 (nitrogen dioxide). These may be free or associated with either water or solid particles in the atmosphere.

Reduced nitrogen
Ammonia is the main form, but various organic compounds may be present and even be locally abundant – consider, for example, the aroma around fish processing plants or when pig slurry is put on to fields or the smell of rotting invertebrates at low tide on a warm day.

Water

All the above nitrogen-containing gases may be present in solution, often at saturation, but sometimes at sub- or super-saturated concentrations. Because the volume of water, particularly in oceans, is so great, its total content of nitrogen-containing dissolved gases is greatly in excess of that in the atmosphere. Water may also contain low concentrations of urea, ammonia and low molecular mass organic compounds such as amino acids.

Living organisms
Nitrogen gas

This will always be present in solution in the cell, normally in equilibrium with the surrounding atmosphere. Possible exceptions include animals moving rapidly with respect to changing outside pressure – seen when divers suffer from the 'bends' when they rise from depth to the surface too quickly.

Oxidized nitrogen

Nitrate may be present in various concentrations in plants, micro-organisms and the alimentary systems of herbivores and omnivores, including man.

Reduced nitrogen

Ammonia may be present in low concentrations in all organisms. At physiological pH values most of it is normally protonated, since the pK of the reaction

$$NH_3 + H_2O \rightleftharpoons NH_4^+ + OH^-$$

is 9.25 at 25 °C (ranging from 10.08 at 0 °C to 8.56 at 50 °C). Organic reduced nitrogen occurs in many forms, including urea, organic bases such as purines and pyrimidines, amino compounds and their polymers (peptides, proteins), polymers of amino sugars (e.g. chitin in exoskeletons of some arthropods and fungal cell walls, cell walls of bacteria) and various secondary products made by plants for defence against herbivores and other problems. All animals have nitrogenous excretory products (Table 1.1) and these will contribute to any of the environmental pools (water, air, soil) of nitrogen.

Table 1.1. *Nitrogenous compounds excreted by major groups of animals*

Compound	Animal group
Ammonia	Aquatic invertebrates, including some molluscs
	Some fish, especially those living in fresh water
	Amphibians in water
Trimethylamine oxide	Some marine teleosts (bony fish)
Urea	Some molluscs
	Elasmobranchs (cartilaginous fish)
	Some amphibians
	Mammals
Purines, for example uric acid	Some terrestrial molluscs
	Insects
	Spiders
	Reptiles
	Birds

Soil

Since soil has a solid, a liquid and a gas phase, all of the substances listed above may be present. In addition, vast amounts of N_2 are occluded in rocks. Ammonium is the principal form of inorganic nitrogen in many undisturbed soils: it may exchange with cations on soil colloids in a reversible way or from more stable complexes which account for up to 5% of total nitrogen in soil (Smith, 1982b). The exchangeable ammonium continually replenishes that taken from the soil solution by microorganisms and plants. Further losses from soil may occur due to leaching (generally small, because of the complexing with soil colloids) or due to volatilization (generally small, except under special, usually alkaline, conditions). A further fraction, known as humic acid, is unique to soils. Humic acid is a generic term used to cover a wide range of complex substances formed by the condensation of aromatic compounds (degradation products of lignin and tannin; flavonoids) with proteins or amino acids. The molecular mass of humic acid varies greatly (from less than 1 kilodalton to several hundred kilodaltons) as does the difficulty with which it is broken down by microorganisms. Large quantities of soil nitrogen may be in this form.

The processes of the nitrogen cycle

It is possible to construct diagrams of the nitrogen cycle from many standpoints, emphasizing, for example, the organisms or the

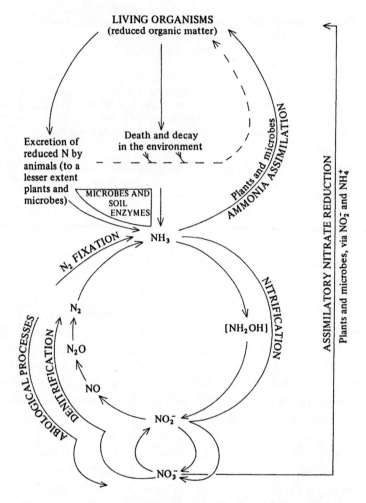

Figure 1.1. A summary of the major reactions of the nitrogen cycle.

chemistry. Figure 1.1 is yet another form, in which the cycle is considered as three interlocking component cycles with a major bypass. The upper cycle is the only one which can operate on its own, for example in some forest climax situations. The central cycle can only operate in association with the upper part, since the organisms which carry out its key reactions must contain organic nitrogen. The smallest section, which contains only nitrate and nitrite, usually operates with the other two parts, but at least in theory, may operate without the central part. The route from nitrate to plant or microbial biomass is shown as a separate branch, since, although nitrite and ammonia are intermediates, the whole process occurs within individual cells in a series of coupled reactions. A dissimilatory pathway from NO_3^- to NH_3 (see pp. 25 and 43) is not included in the figure, as details are unclear.

Figure 1.1 is purely descriptive, giving no indication of the total quantities of the different components nor of the energy contained within them. The former will be considered in the next chapter. In order to know whether reactions are likely to occur, it is necessary to know how much energy is required by, or lost during, those reactions. Table 1.2 gives values for the major reactions of the nitrogen cycle. The actual energetics are more complex for a number of reasons. First, the rate of reactions must be considered – if a reaction is thermodynamically possible, but occurs very slowly, it is generally of little use to living organisms. Biological catalysts (enzymes) are the common way of speeding up reactions: their synthesis and maintenance require energy. Additionally, some of the available energy may be used to drive the reaction. Second, where reactions are required to occur at low substrate concentration (for example if little substrate is available, or if the substrate is toxic) it may be more advantageous (competitive) to use energy (usually as adenosine triphosphate, ATP) to help drive that reaction, so that a more rapid reaction rate is achieved. An example of this in the nitrogen cycle is the assimilation of ammonium by the GS: GOGAT system, discussed below. Third, the partial reactions may have different energies associated with them, so that an overall reaction which is energetically feasible may not occur at finite rates in practice. The classic example of this in the nitrogen cycle is the reduction of nitrogen gas to ammonia. Fourth, living organisms have various potential energy sources available to them and these sources vary considerably in the fraction of their energy available to drive reactions. This means, for example, that the actual energy used by organisms to fix nitrogen varies greatly. Fifth, energy available for oxidation–reduction reactions (i.e. all the reactions of the nitrogen cycle) will vary according to whether the environment is oxidative or reducing. Some of these aspects are considered later in this chapter. Quantitative details are given in Harris (1982).

Each of the steps in the cycle is subject to constraints related to the physical and biological environment. Thus the actual pathway followed and its rate-limiting steps, vary with geographical location and season. The purpose of this book is to explore these variations, but first it is necessary to look at the reactions themselves.

Table 1.2. *Standard free energy changes ($\Delta G_0'$) associated with some of the major reactions of the nitrogen cycle. A negative sign indicates net energy loss. More details of individual reactions can be found in Table 1.3. From various sources including Harris (1982) and Rosswall (1982)*

Reaction	$\Delta G_0'$ (kJ mol^{-1})	Comment
1. $NO_3^- \rightarrow NO_2^-$	-161	Nitrate respiration, anaerobic
2. $NO_3^- \rightarrow NO_2^-$	-142 to -161	Assimilatory nitrate reduction: value depends on electron donor (see Table 1.3)
3. $NO_2^- \rightarrow NH_4^+$	-374 to -433	Anaerobic nitrite reduction: value depends on electron donor (see Table 1.3)
4. $NO_3^- \rightarrow NH_4^+$	$+348$	The sum (2 + 3) taking into account the energy required to produce the electron donor (see also Sprent & Raven, 1985)
5. $\frac{1}{2}N_2 \rightarrow NH_4^+$	-40	Theoretical value for overall nitrogen fixation reaction. Initial step requires energy (see text)
	$+480$	Taking into account energy needed to produce electron donors, wasted in H_2 evolution, etc. (see Sprent & Raven, 1985)
6a. $NH_4^+ \rightarrow NH_2OH$	$+15$	⎫ As carried out by
6b. $NH_2OH \rightarrow NO_2^-$	-244 to -353	⎬ *Nitrosomonas*
7. $NO_2^- \rightarrow NO_3^-$	-65 to -88	As carried out by *Nitrobacter*
8. $N_2O^- \rightarrow N_2O$	-230	
9. $NO \rightarrow \frac{1}{2}N_2O$	-153	
10. $NO_2^- \rightarrow NO$	-76	
11. $\frac{1}{2}N_2O \rightarrow \frac{1}{2}N_2$	-170	
12. $NO_3^- \rightarrow \frac{1}{2}N_2$	-560	An overall figure, based on denitrification by *Pseudomonas aeruginosa*
13. $NO_3^- \rightarrow NH_4^+$	-591	Nitrate fermentation (dissimilatory), anaerobic: as carried out by *Clostridium perfringens*

Abiological processes

Nitrogen fixation can occur naturally, or as a result of industrial reactions. In either case, because of the strength of the nitrogen–nitrogen bond, energy is required. Suitable energy sources are short wavelength light (photochemical reactions) and electric discharge (thunderstorms), either of which may be sufficient to cause nitrogen and oxygen to combine, forming one or more of the nitrogen oxides described above. Industrially, the Haber–Bosch process in which nitrogen and hydrogen are combined under pressure and at high temperature (these together providing the necessary energy) to form ammonia is the most widespread way of producing nitrogen fertilizer. Free ammonia may be used or it may be further processed to ammonium salts or urea. Unintentional industrial 'fixation' of nitrogen occurs during various high-temperature combustion processes. According to the fuel (coal, wood, diesel oil, aviation spirit, etc.), the temperature and pressure of the combustion, and the rate of air supply, the products may be nitrogen oxides, ammonia or even hydrogen cyanide. The quantities of nitrogen involved in this component of the abiological cycle have increased greatly in recent years and because of further reactions which may result there is cause for concern about possible adverse effects on the environment. This topic will be discussed in chapter 7.

In addition to that from industrial processes, a considerable amount of nitrogen, both free and combined, is passed to the atmosphere from natural (wild) or man-made fires (for example burning of straw).

Enzymic reactions
Most of these occur in living organisms, but some important reactions are catalysed by soil enzymes, of which urease is the most important in the context of nitrogen cycling. Table 1.3 lists the major enzymes of the nitrogen cycle.

Biological nitrogen fixation
The enzyme complex known as nitrogenase is found only in certain microorganisms. These were formerly known as prokaryotes, but have recently been divided into two kingdoms, prokaryotes and archaebacteria (also known as metabacteria), see, for example, Stackebrandt & Woese (1984): the occurrence of nitrogenase in eukaryotes has been reported from time to time, but none of these reports has been confirmed. Nitrogenase catalyses two simultaneous and, in living organisms, inseparable reductions, those of dinitrogen gas and protons:

$$8H^+ + N_2 + 8e^- \rightarrow 2NH_3 + H_2$$

Note that, for every nitrogen molecule reduced, one molecule of hydrogen is produced; under some circumstances, considerably more hydrogen is formed. The details of the reactions are very complex and not fully

Table 1.3. *Major enzymes of the nitrogen cycle. See also Figure 1.2 and p. 19*

EC number	Systematic name (recommended trivial name)	Reaction	Further comments	Table 1.2 reaction number
1.6.6.1	NADH:nitrate oxidoreductase (nitrate reductase (NADH))	$NADH + NO_3^- \rightarrow NAD^+ + NO_2^- + H_2O$	NADH specific	2
1.6.6.2	NAD(P)H:nitrate oxidoreductase (nitrate reductase (NAD(P)H))	$NAD(P)H + NO_3^- \rightarrow NAD(P)^+ + NO_2^- + H_2O$	Uses NADH or NADPH	
1.6.6.3	NADP:nitrate oxidoreductase (nitrate reductase (NADPH))	$NADPH + NO_3^- \rightarrow NADP^+ + NO_2^- + H_2O$	NADPH specific	
			Eukaryotic ‑ Contain Mo and flavin	
1.7.7.2	Fd_{red}:nitrate oxidoreductase (nitrate reductase (Fd))	$Fd_{red} + NO_3^- \rightarrow Fd_{ox} + NO_2^- + H_2O$	Ferredoxin specific, prokaryotic	1
1.7.99.4	Respiratory nitrate reductase (nitrate reductase)	$acc_{red} + NO_3^- \rightarrow acc_{ox} + NO_2^- + H_2O$	Contains Mo	1
1.6.6.4	NAD(P)H:nitrite oxidoreductase (nitrite reductase (NAD(P)H))	$3NAD(P)H + NO_2^- \rightarrow 3NAD(P)^+ + NH_4OH + H_2O$	Found in non-photosynthetic organisms	3
1.7.7.1	Ammonia:ferredoxin oxidoreductase (ferredoxin nitrite reductase)	$6Fd_{red} + NO_2^- \rightarrow 6Fd_{ox} + NH_4OH + H_2O$	Characteristic of photosynthetic organisms	3
1.7.99.3	Nitric oxide (acceptor):oxidoreductase (nitrite reductase)	$2NO + 2H_2O + acc_{ox} \rightarrow 2NO_2^- + acc_{red}$	Contains Cu	10
1.7.99.2	Nitrogen (acceptor):oxidoreductase (nitric oxide reductase)	$N_2 + acc_{ox} \rightarrow 2NO + acc_{red}$		
1.7.3.4	Hydroxylamine:oxygen oxidoreductase (hydroxylamine oxidase)	$NH_2OH + O_2 \rightarrow NO_2^- + H_2O$		6b
1.18.2.1	Terminology not universally agreed (nitrogenase)	$N_2 + 8H^+ + 8e^- \rightarrow 2NH_3 + H_2$	Consists of two components, sometimes considered to be separate enzymes (EC 1.18.6.1 and 1.19.6.1), nitrogenase reductase (=Fe protein or component 2) and dinitrogenase (=Mo–Fe protein or component 1). The electron donor in most systems is probably reduced ferredoxin	5
None allocated	Not fully characterized (nitrous oxide reductase)	$N_2O + 2H \rightarrow N_2 + H_2O$	Contains Cu. Reduced cytochrome c' may act as electron donor in some spp.	11
3.5.1.5	Urea amido hydrolase (urease)	$NH_2CONH_2 + H_2O \rightarrow CO_2 + 2NH_3$	Contains Ni	

understood, but a concise account of the current situation is given in Postgate (1987). The actual reduction of nitrogen occurs at a site on the enzyme where molybdenum is found. Breaking of the nitrogen to nitrogen bond appears to be the major stumbling block. To illustrate this, we shall assume that the reduction occurs in a three-step process as follows (Harris, 1982):

$$H_2 + N_2 \rightarrow N_2H_2 \qquad\qquad +209.2 \text{ kJ mol}^{-1}$$
$$H_2 + N_2H_2 + H^+ \rightarrow N_2H_5^+ \qquad -86.7 \text{ kJ mol}^{-1}$$
$$H_2 + N_2H_5^+ + H^+ \rightarrow 2NH_4^+ \qquad -201.5 \text{ kJ mol}^{-1}$$
$$\text{overall reaction} \qquad\qquad -79.0 \text{ kJ mol}^{-1}$$

Thus, although the overall process is exergonic, the initial step is endergonic. The nitrogenase reaction does not proceed in exactly this way, but the energy requirement for the initial step is undisputed. Two ATP molecules are thought necessary to provide the energy to transfer each electron. This energy is usually derived ultimately from photosynthesis, but occasionally from chemosynthesis. Some nitrogen-fixing organisms can photosynthesize, either by the higher plant (oxygen-evolving) pathway (cyanobacteria or blue–green algae), or by 'bacterial photosynthesis', which does not involve oxygen evolution, because it uses compounds other than H_2O (e.g. H_2S) to provide the reducing power required for the synthesis of organic compounds from carbon dioxide. These latter organisms do not have the potential problem of oxygen inactivation of nitrogenase, a problem posing a considerable dilemma for organisms which either need oxygen to produce ATP (by respiration) or generate it in photosynthesis. Ways in which this problem may be overcome are listed in Table 1.4 and are further discussed in Sprent (1979), Postgate (1982) and Stewart & Rowell (1986). Tables 1.5 and 1.6 list the major genera of cyanobacteria and 'other' bacteria which fix nitrogen in the free-living state.

Nitrogen-fixing organisms which are not autotrophic need a source of reduced carbon on which to grow. This may be obtained from the environment (water, soil (including peat), root exudates, etc). If combined nitrogen is also present, the organisms use this (because it requires less energy to assimilate). They are then in competition with other organisms using combined nitrogen. Alternatively, carbon may be supplied by a host eukaryotic organism – usually a plant, but occasionally an animal living on a high C : N diet (termites, Prestwick & Bentley, 1981; shipworms, Waterbury, Calloway & Turner, 1983). It is the varying ultimate sources of energy used to drive the nitrogenase reaction, plus the energy needed in some organisms to protect the nitrogenase from oxygen,

Table 1.4. *Some ways in which oxygen inactivation of nitrogenase may be avoided*

Method of avoidance	Examples
Strict anaerobiosis, at least during N$_2$ fixation.	*Clostridium, Klebsiella*
N$_2$ fixation only when pO$_2$ is low or zero	Some non-heterocystous cyanobacteria, e.g. *Trichodesmium*
	Some bacteria, e.g. *Azospirillum*
Use of energy from non-O$_2$-evolving photosynthesis	*Rhodospirillum*
Spatial separation of O$_2$-evolving photosynthesis from N$_2$ fixation	Heterocystous cyanobacteria, e.g. *Nostoc, Anabaena*. Heterocysts can carry out photophosphorylation but do not reduce CO$_2$ or evolve O$_2$
Temporal separation of N$_2$ fixation from O$_2$-evolving photosynthesis	Some unicellular cyanobacteria, e.g. *Gloeothece*
Modified respiratory pathways consuming excess O$_2$. Not coupled with ATP synthesis	*Azotobacter*
Conformational protection of nitrogenase, which is temporarily inactive	*Azotobacter*
Physical diffusion barriers $\pm$ coupled with O$_2$ transport involving haemoglobin. Overall a high flux of O$_2$ at a low concentration is maintained	Nodules on legumes. Actinorhizas

that results in the overall efficiency (energy used per nitrogen molecule reduced) varying greatly with both organism and environmental conditions (Sprent, 1979; Harris, 1982).

Most of the nitrogen fixation taking place in present times occurs in symbiotic systems, in particular those legumes (and one non-legume) which are nodulated by *Bradyrhizobium* or *Rhizobium* and the scattered non-legume genera nodulated by *Frankia* (usually known as actinorhizas). Most widely studied have been the agriculturally important grain and forage legumes, but increasing attention is now being paid to the ecologically important actinorhizas and woody legumes. Table 1.7 and 1.8 summarize the extant legumes and non-legumes known to fix nitrogen. Cyanobacterial symbioses occur in all major plant groups: these are summarized in Table 1.9. The general biology of most of these nitrogen-fixing organisms is discussed in Sprent (1979) and Postgate (1982), with some more recent material (including newly discovered

Table 1.5. *Major genera of cyanobacteria which are known to fix nitrogen in particular habitats. H indicates heterocystous. For details see Fay (1981) and Postgate (1982)*

Habitat	Genus	Comment
Terrestrial		
	Anabaena H	
	Anabaenopsis H	
	Aulosira H	Found from poles to equator.
	Calothrix H	Fixation only occurs in wet
	Cylindrospermum H	conditions, but some can
	Fremyella H	survive prolonged desiccation
	Gloeotrichia	in a virtually inactive state.
	Hapalosiphon H	Important colonizers, e.g. of
	Nostoc H	volcanic lava soil within 1.5 y of
	Scytonema H	an eruption
	Stigonema H	
	Tolypothrix H	
Fresh water		
	Anabaena H	Forms dense planktonic mats.
	Aphanizomenon H	Gas vacuoles permit adjust-
	Gloeotrichia	ment of vertical distribution
	Calothrix H	May be important in hot springs
	Mastiglocladus	
	Gloeothece	Unicellular, only some strains fix
	Synechococcus	nitrogen and rarely aerobically (Huang & Chow, 1986)
Marine		
	Oscillatoria	Also known as *Trichodesmium*. May be important in open sea
	Calothrix H	
	Dichothrix H	
	Lyngbia	
	Nostoc H	
	Oscillatoria	Particularly important in inter-
	Rivularia H	tidal regions of coral reefs
	Schizothrix	
	Scytonema H	
	Tolypothrix H	
Marine epiphytes		
	Calothrix H	Common on seagrasses
	Dichothrix H	Common on *Sargassum*
	Rivularia H	

Table 1.6. *Principal bacterial genera with species or strains of species which fix nitrogen in the free-living state. For cyanobacteria, see Table 1.5. Amongst the heterotrophs only those marked * regularly fix nitrogen gas*

Nutritional type	Genus	Comment
Phototrophs: anaerobic	*Amoebobacter*	
	Chlorobium *Chromatium*	} Moist and muddy soils, fresh and salt water fixation rare. Sulphur bacteria
	Ectothiorhodospira *Pelodictyon* *Rhodomicrobium* *Rhodopseudomonas*	
	Rhodospirillum	} Widely studied. Light required for N_2 fixation
	Thiocapsa *Thiocystis*	
Heterotrophs	*Alcaligines* *Aquaspirillum* *Arthrobacter* *Azomonas* *Azospirillum* * *Azotobacter* * *Azotococcus* * *Bacillus* *Beijerinckia* * *Campylobacter* *Citrobacter*	
	Clostridium	} Widely studied, particularly biochemically
	Derxia * *Desulfotomaculum* *Desulfovibrio* *Enterobacter* *Erwinia* *Escherichia*	
	Klebsiella	} Widely studied, particularly for genetics and biochemistry
	Methylobacter *Methylococcus* *Methylocystis* *Methylomonas* *Methylosinus* *Thiobacillus* *Xanthobacter*	

Table 1.7. *Summary of the sections of the Leguminosae which have nitrogen-fixing root nodules. Classification according to Polhill & Raven (1981). For details see Allen & Allen (1981); Faria et al. (1987); Sprent, Sutherland & Faria (1987b).*
Note that more than half the known legume genera have not been checked for nodulation. Recent evidence suggests that many are not infected by the well-known root hair process (Sprent et al., *1987b*)

Sub-family	Tribe	Comment
Papilinoideae	All tribes except where noted	The Dipterygeae is the only tribe with negative reports in all (3) genera. Isolated genera in other tribes do not appear to nodulate
Mimosoidseae	All tribes	Nodulation may vary within genera. For example, not all *Acacia* spp. nodulate
Caesalpiniodeae		Nodules have unique structural features
	Cassieae	Only species placed in (or closely related to) the genus *Chaemacrista* nodulate
	Caesalpinieae	Confirmed reports only on 7 genera
	Detarieae	7 of the 8 positive reports need further confirmation
	Amherstieae	1 report, needs confirmation

types of nodule structure) in Sprent, Sutherland & Faria (1987a). References to specific systems will be given in later chapters. Nitrogen-fixing organisms are particularly important (a) in agriculture and forestry, to replenish nitrogen removed with the crop plant or animal; (b) on pioneering sites where there is little combined nitrogen, for example sand dunes, volcanic debris; (c) to replace losses due to fire or denitrification.

Ammonium and its assimilation by organisms

Like nitrate, ammonium cannot be assimilated by animals, except those with a suitable gut flora. Plants and microorganisms can carry out the process by one of two main routes: first, the glutamine synthetase: glutamate synthase system, commonly known as GS: GOGAT (Figure 1.2A). Because of the low $K_{1/2}$ (the currently preferrred term for what was formerly known as the Michaelis constant, K_m) of GS for ammonium and the general toxicity of ammonium to cells making rapid assimilation

Table 1.8. *Genera of nodulated non-leguminous plants. In all except* Parasponia, *which is infected by* Rhizobium, *the endophyte is the Actinomycete,* Frankia *spp.*

Family	Genus	Comment
Betulaceae	*Alnus*	All spp. probably nodulate
Casuarinaceae	*Allocasuarina*	The genus *Casuarina* has now been divided into 4, only 3 of which are so far named. Not all spp. appear to nodulate
	Casuarina	
	Gymnostoma	
	unnamed genus	
Coriariaceae	*Colletia*	
	Coriaria	All spp. probably nodulate
Datiscaceae	*Datisca*	
Eleagnaceae	*Eleagnus*	Infection via epidermis, not root hairs
	Hippophäe	
	Shepherdia	All spp. probably nodulate
Myricaceae	*Comptonia*	Sometimes included in *Myrica*
	Myrica	All spp. probably nodulate
Rhamnaceae	*Ceanothus*	All spp. probably nodulate
	Discaria	
	Kentrothamnus	
	Retamnilla	
	Talguena	
	Trevoa	
Rosaceae	*Cercocarpus*	
	Chamaebatia	
	Cowania	
	Dryas	All spp. probably potentially nodulated, but nodules not always found at all locations
	Purshia	
Ulmaceae	*Parasponia*	The only recorded non-legume infected with *Rhizobium*. Not infected via root hairs

desirable, this is thought to be the major route of ammonium assimilation, in spite of its ATP requirement. Second, the glutamate dehydrogenase system (Figure 1.2B). This reversible reaction has a higher $K_{1/2}$ for ammonium than GS and is thought to be more significant in the degradative direction. These matters are discussed in standard biochemistry texts, as are the reactions for synthesizing other organic nitrogen compounds.

Breakdown of organic material following death

Organisms vary from large (elephants, trees) to very small (microorganisms). Size itself can be a problem in the process of mineralization

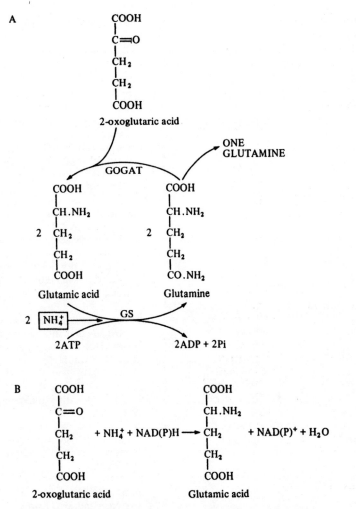

Figure 1.2. Reactions which may incorporate ammonium into amino acids. A, the glutamine synthetase (GS) : glutamate synthase (GOGAT = glutamine-2-oxoglutarate-amino-transferase) system. Note that for every $2NH_4^+$, one glutamine is the net product. B, the glutamate dehydrogenase system.

(also known as ammonification, the conversion of complex organic nitrogen to ammonium). Large animals are generally processed into smaller units by a range of other animals and microorganisms. The wood of large trees is initially broken down mainly by fungi, but the leaves and smaller pieces of wood may be consumed by invertebrates, ranging from earthworms in moister environments to various arthropods in drier areas. In both cases decaying material is usually taken underground. The vital

Table 1.9. *Nitrogen-fixing symbioses in which the microsymbiont is cyanobacterial.*

Macrosymbiont Division	Genus	Comment
Chrysophyta	*Chaetoceros* *Rhizosolenia*	Both these diatoms may contain the heterocystous species *Richelia intracellularis*
Anthocerotophyta Hepaticophyta	*Anthoceros* *Blasia* *Cavicularia*	*Nostoc* sp. inhabits cavities on underside of thalloid gametophyte. N_2 fixation supported by host photosynthate
Pteridophyta	*Azolla*	*Anabaena azollae* inhabits cavities on the underside of the sporophyte leaf. N_2 fixation supported by host photosynthate. Economically important as fertilizer for flooded rice crops and for potential blocking of waterways
Cycadophyta	*Bowenia* *Ceratozamia* *Cycas* *Dioon* *Encephalartos* *Lepidozamia* *Macrozamia* *Microcycas* *Stangeria* *Zamia*	All genera appear to contain *Nostoc* or *Anabaena* in modified coralloid roots which may be superficial or subterranean. N_2 fixation supported by host photosynthate. Probably very important 150 My ago. Now locally significant for N_2 fixation
Angiospermae	*Gunnera*	The only known genus of flowering plant with a cyanobacterial symbiont
Lichens	Numerous genera	All lichens with a cyanobacterial partner fix N_2. In two-membered forms (fungus + cyanobacterium) the cyanobacterium also fixes CO_2. In three-membered forms photosynthesis is probably a function solely of the algal member

role of protozoa amongst the microorganisms carrying out the later stages of breakdown is now being appreciated (see, for example, Clarholm, 1985). Mineralization reactions are driven by the need of microorganisms for carbon, and release of ammonium eventually follows (see next chapter). In addition to microbiological reactions, when particle size has been sufficiently reduced, free enzymes may also be involved. These originate in microorganisms either by secretion or following cell death. In the presence of suitable clay particles these enzymes (and other proteins) may be strongly adsorbed, rendering them resistant to microbial attack. Degradation processes in soil may be continued by such enzymes in the absence of living organisms. For a general account of soil enzymes see Burns (1978). One of the most widely studied is the hydrolytic enzyme urease, which is much more stable in soil than under normal laboratory conditions! Its substrate, urea, may be present in soil or water as an animal excretory product, and in fertilizer, or be produced from organic substances such as purines (including uric acid and similar excretory products). Under some conditions urease activity can lead to a considerable loss of ammonia from soils – this may be a major limitation to the use of urea as a fertilizer and has led to a search for urease inhibitors suitable for use in soils (chapter 7). Urease may also occur in high concentrations in plants (particularly some grain legumes such as soybeans, *Glycine max*). Such plants can take up urea directly from soil. It is possible that many other soluble organic compounds (especially amino acids) can be utilized directly by plants and microorganisms, thus short-circuiting parts of the nitrogen cycle.

It is during the process of decay that formation of humic acid occurs (see p. 5): this may essentially immobilize large fractions of organic nitrogen in soil. Thus it is possible to have soils very rich in total nitrogen, but very low in available nitrogen: peat soils are a classic example. In these, breakdown is further hampered by a high C:N ratio (pp. 31–4).

From ammonium to nitrate (nitrification)

The major pathway (Figure 1.3) is carried out by the two widespread gram-negative, chemoautotrophic genera *Nitrosomonas* and *Nitrobacter* and some closely related species (Table 1.10). Both carry out nitrifying reactions aerobically, but may be able to reverse them under anaerobic conditions. The initial reaction, catalysed by ammonia monooxygenase is $NH_3 + O_2 + 2[H] \rightarrow 2NH_2OH + H_2O$. The exact nature of the reductant is not known (Wood, 1986). Because the numbers of nitrifying bacteria in natural environments are greatly exceeded by heterotrophic species, enrichment, isolation and study of them tend to be tedious processes. Growth of both *Nitrosomonas* and *Nitrobacter* is slow with natural generation times probably about 20–40 hours. This and the small numbers of cells may give a misleading impression of their importance to the nitrogen cycle. The apparent paradox is explained by a consideration of the energetics of using the oxidation of ammonia or nitrite for growth (see

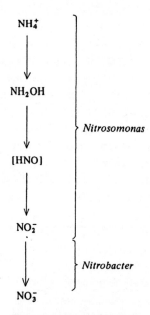

Figure 1.3. The major pathway from ammonium to nitrate. Hydroxylamine (NH_2OH) is generally considered to be an intermediate. For details, see the volume edited by Prosser (1986).

Table 1.2). The energy from such oxidation is all that is available both for the fixation of carbon dioxide and all other reactions needed to make cells. Carbon dioxide fixation, even if 100% efficient, requires at least 480 kJ of energy per mole. Biological processes (and industrial ones) tend to proceed at much less than maximum efficiency – a necessary feature if they are to achieve reasonable *rates*, efficiency and speed being to some extent traded against each other (this is another way of saying that some of the available energy is used to drive the reaction).

Oxidation of nitrite is almost certainly accompanied by electron flow through the cytochrome (a + a_3) complex to oxygen: the oxygen incorporated into the nitrate is derived from water. The redox potential of the cytochrome (a + a_3) – oxygen couple is much higher (less negative) than that associated with the generation of NADPH, the reducing power which is used for most of the synthetic reactions of growth, including carbon dioxide fixation. In order, therefore, to make NADPH, it is necessary to use energy from the hydrolysis of ATP to reverse the normal electron flow pathway. Overall then, it is not surprising that *Nitrosomonas*

Table 1.10. *Genera of nitrifying bacteria: all are in the family Nitrobacteriaceae. Carbon dioxide is fixed by the Calvin cycle, using reducing power from the oxidation of ammonium or nitrite. Further details in Watson, Valois & Waterbury (1981), Sharmer & Ahler (1977), Prosser (1986)*

Metabolic grouping	Genus and species	Morphology and anatomy	Comment
Ammonium oxidizing	*Nitrosomonas europaea*	Usually separate rods with rounded ends: often aggregated	Only 1 sp. now recognized, but much strain variation especially in relation to habitat (marine, freshwater or soil)
	Nitrosovibrio tenuis	Curved rods: extensive cytomembrane system	Only one strain known
	Nitrosococcus mobilis	No extra wall layers	One strain: obligatorily marine
	Nitrosococcus nitrosus	Large, spherical	No obligate NaCl requirement. Laboratory culture now lost!
	Nitrosococcus oceanus	Large spheres with a stack of central, flattened vesicles and 2 extra wall layers	Obligatorily marine. Uncommon
	Nitrospira briensis	3–20 turn spiral. Motile or non-motile	Soils
	Nitrosolobus multiformis	Lobed with large amd smaller compartments. Motile	Widespread in soils
Nitrite oxidizing	*Nitrobacter winogradskyi*	Rods which bud and have unique cell envelope	Soil, marine and freshwater. Some grow heterotrophically on acetate
	Nitrococcus mobilis	Spherical, motile with tubular cytomembrane	Only one (marine) strain isolated
	Nitrospira gracilis	Long slender rods with extensive cytomembrane	Only one strain so far

A

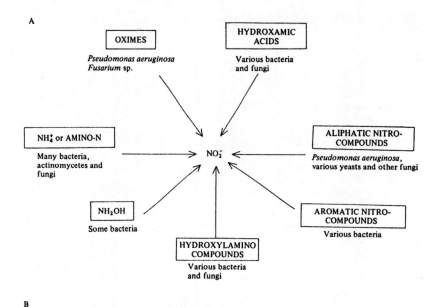

B

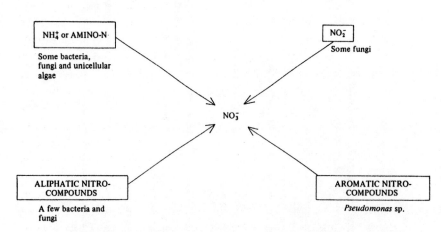

Figure 1.4. Minor pathways to nitrite (A) or nitrate (B). The reactions are carried out with heterotrophic organisms and are not usually linked to cell growth.

uses about 35 NH_4^+ and *Nitrobacter* about 100 NO_2^- per CO_2 fixed. Or, to put it another way, a comparatively few cells can process a large number of molecules of combined nitrogen. Note that in the overall reaction leading from ammonium to nitrate $2H^+$ are generated per N; this may cause acidification in some environments (Kennedy, 1986).

In soils and in marine environments, *Nitrosomonas* and *Nitrobacter* together carry out most of the nitrification which occurs. However, there is a range of organisms capable of producing both nitrate and nitrite heterotrophically. These organisms include bacteria, fungi, and some unicellular algae. Figures 1.4 A and B summarize the reactions, which are discussed in detail by Focht & Verstraete (1977). Some fungi actually accumulate nitrate. Generally the reactions outlined in Figure 1.4 are not used to support growth: instead they are used to inactivate toxins (including nitrite), to metabolize substances which may be required for the growth of other, competing organisms and probably many other reasons associated with life in the highly competitive microbial world. These include acquisition of trace elements, since products such as hydroxamic acids are powerful iron chelators. Heterotrophic nitrifying organisms may be important in nitrogen cycling in soils of extreme pH, both acid and alkaline.

Pathways from nitrate and nitrite

Nitrate is the most abundant form of combined nitrogen in most agricultural soils, most desert soils and most waters. Thus it is not surprising that many organisms can utilize it. However, these include no animals, only plants and microorganisms (eubacteria, archaebacteria and fungi). Figure 1.5 summarizes the pathways beginning at nitrate. The first step is common, reduction of nitrate to nitrite, catalysed by nitrate reductase. It may be linked to growth (assimilatory nitrate reduction) or not (dissimilatory nitrate reduction): the latter is also known as nitrate respiration when nitrate is used instead of oxygen as a terminal electron acceptor. At least 73 bacterial genera, including nitrogen-fixing types such as *Rhizobium* and *Azospirillum* can carry out dissimilatory nitrate reduction – a comprehensive list can be found in Jeter & Ingraham (1981) – as can some plants. The nitrite produced can be utilized by a variety of organisms, including denitrifying species: there are about 25 genera of these, ranging from photoautotrophs to human pathogens. They produce N_2O or N_2 gas. True denitrification, like nitrogen fixation, does not occur in eukaryotes.

Reactions B, C, and D as well as A may be coupled to ATP production, although there are some grave doubts about this occurring at the nitrite reductase step. Therefore, in theory at least, growth should be possible on any one step in the sequence, so it is not surprising that organisms are known which are genetically unable to carry out the whole sequence of reactions leading to N_2 gas. There may also be physiological reasons (lack

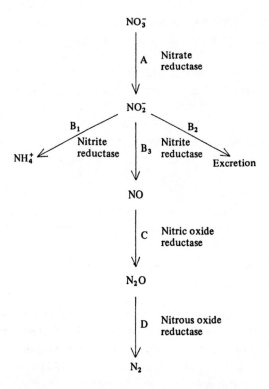

Figure 1.5. The major pathways from nitrate. B_1 leads to amino acid synthesis (see Figure 1.2); B_2 may be coupled to ATP synthesis when nitrate is used as a terminal electron acceptor; B_3 leads to denitrification. There is considerable doubt about nitric oxide as an intermediate: alternatives are given in Figure 1.6. ATP synthesis in D may be linked to generation of a membrane potential (McEwan *et al.*, 1985).

of substrate for example) why certain reactions are absent from a particular system. Evolution of gas (N_2, N_2O) is generally taken as a criterion for denitrification. However, because N_2O is more soluble than N_2, lack of gas evolution does not necessarily mean that denitrification reactions are absent.

Denitrification occurs in almost all known environments (Table 1.11), from arctic to antarctic and including hot springs. It is particularly prevalent in sewage, heavily fertilized soils and other environments which are rich in combined nitrogen. Many organisms, including again some nitrogen-fixing species, can carry out the various reactions. Indeed, most organisms involved in the nitrogen cycle can, under appropriate circumstances, use their key reactions in either direction. The nitrogen-

Table 1.11. *Habitats which may contain denitrifying organisms: summarized from Jeter & Ingraham (1981). Many organisms are found in more than one type of environment*

Fresh water
Salt water, including brines used industrially
Soil
Compost
Dead animal tissue
Live animal tissue – pathogens
                   – non-pathogens
Dead plant tissue
Live plant tissue – pathogens
                  – non-pathogens
Food (human and other animal)
Sewage

fixing species are rather different in this respect in that nitrogen is fixed and produced by entirely different reactions, *both* being reductions. The general biology of denitrifying bacteria is discussed by Jeter & Ingraham (1981) and their chemistry and physiology amplified in the monograph by Payne (1981). A recent account of their biochemistry is given by Fillery (1983): for general reviews see Firestone (1982) and Knowles (1982).

Denitrification takes place under anaerobic conditions. However, most denitrifying organisms can also live aerobically. Their metabolic regulation is probably effected by oxygen, its absence being necessary for synthesis of denitrifying enzymes. Even in drained soils, there may be many anaerobic microsites which can support denitrifying bacteria. Given the diverse genera involved, the reader may guess that a wide variety of substrates can be used for growth of denitrifying bacteria: some are listed in Table 1.12. Complete reduction of $2NO_3^-$ to N_2 generates $2OH^-$, which may cause environmental pH to rise. Recent evidence suggests that some bacteria can denitrify aerobically (Lloyd, Boddy & Davies, 1987).

Some bacteria can reduce nitrate to ammonium by a dissimilatory pathway, i.e. one where it is not incorporated into cells (Yordy & Ruoff, 1981; Rosswall, 1982). This, energetically, is a very favourable way of processing any nitrate which may be formed in, or be carried to, anoxic environments (see also p. 43).

Nitrite may also be used in a dissimilatory manner by various microorganisms. These include the marine species *Paracoccus halodenitrificans*

Table 1.12. *Examples of growth substrates for denitrifying bacteria*

Carbon dioxide (phototrophs)
Carbon dioxide + hydrogen
Cyclitols (sugar alcohols)
Methanol
Organic acids, both mono- and dicarboxylic
Phenol
Saccharides
Sulphur
Thiosulphate
Various complex organic substances

(Grant & Hochstein, 1984), the soil bacterium *Citrobacter* sp. (Smith, 1982a) and some photosynthetic green algae (Weathers, 1984) and cyanobacteria (Weathers & Niedzielski, 1986). The products are nitrous oxide and/or ammonium, each of which may be formed under the action of different nitrite reductases.

Nitrite may react abiologically with organic constituents of soil, particularly lignin and its breakdown products. This may yield N_2, N_2O, $NO + NO_2$, and CH_3ONO (methyl nitrite). Soil nitrite levels are not normally sufficiently high for this to contribute significantly to overall denitrification. However, high levels of nitrite may be found in alkaline soils, especially where alkali has been generated from nitrogen-containing fertilizers, such as ammonia and urea (which hydrolyses to form ammonia): when ammonia is protonated, OH^- is generated and pH rises. Calcareous soils given ammonium sulphate are another example. Chemodenitrification leading to the production of nitric oxide may be considerable in environments of low pH (Chalk & Smith, 1983). The exact role of nitric oxide, even as to whether it is normally a free intermediate in biological denitrification, is critically discussed in Fillery (1983). Alternative positions for NO are indicated in Figure 1.6.

Further abiological denitrification processes may occur in the stratosphere from a combination of photodissociation by solar ultraviolet radiation (<260 nm wavelength) (reaction 1), and chemical reactions (2) and (3) (Galbally & Roy, 1983):

$$\text{(1)} \quad N_2O \rightarrow N_2 + O('D)$$
$$\text{(2)} \quad N_2O + O('D) \rightarrow N_2 + O_2$$
$$\text{(3)} \quad N_2O + O('D) \rightarrow 2NO$$

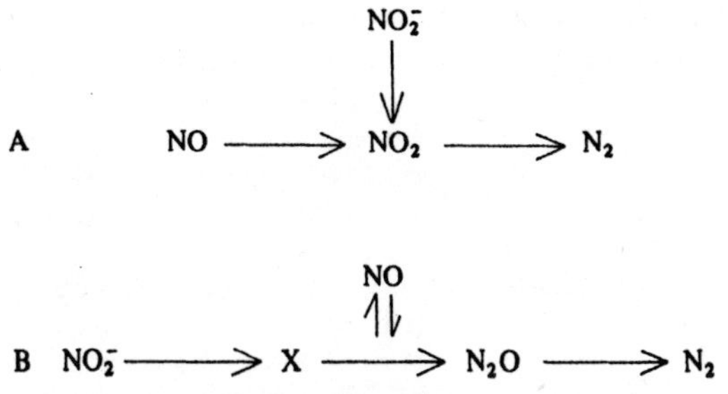

$$A \qquad NO \longrightarrow NO_2 \longrightarrow N_2$$

with NO_2^- feeding into NO_2 from above.

$$B \quad NO_2^- \longrightarrow X \longrightarrow N_2O \longrightarrow N_2$$

with NO in equilibrium above X.

Figure 1.6. Alternative positions for nitric oxide (NO) in denitrification. If not a direct intermediate in denitrification (see Figure 1.5), then NO probably arises by chemodenitrification. X represents an unknown intermediate. For details see Fillery (1983).

where $O('D)$ represents electronically excited atomic oxygen. It is suggested that of the N_2O formed by microbes which is lost from terrestrial ecosystems, some 90% is converted to N_2, the rest in several stages to nitric acid which is returned to the soil and oceans as nitrate. Ozone in the lower atmosphere rapidly converts any NO released from soil into NO_2. This may be absorbed directly (and utilized) by plants (Evans, Canada & Santucci, 1986) or converted to HNO_3 and returned to the surface in rainwater (Galbally & Roy, 1983).

Nitrate to organic nitrogen
In Figure 1.1, this step is given as a bypass to all the other reactions. Many microorganisms and most plants can take up nitrate and incorporate it, after reduction, into organic nitrogen. The enzymes involved are nitrate and nitrite reductases, and the final product is ammonium: this is assimilated by either of the pathways described previously. Any given organism may have more than one form of nitrate reductase (e.g. induced and constitutive enzymes), but chemically these are very similar and always contain molybdenum. Nitrate reduction can be carried out in any organ of higher plants, including root, stem, leaf and fruit. The proportion of the nitrate taken up which is reduced in these different parts varies with species, age and concentration of free nitrate in the environment (Andrews, 1986; Wallace, 1986). The reductant needed to convert nitrate to ammonium is derived either directly or indirectly (via respiration) from photosynthesis. The OH^- generated may be disposed of in soil (root nitrate reduction) or require the production of acid to generate H^+ for pH

Table 1.13. *Versatility of nitrate reduction in plants*

Source of versatility	Component of versatility
Source of reductant	Product of light reaction in leaves
	Respiration (any part of plant including leaves)
Methods of pH regulation (for details see Raven, 1985)	Malate production, followed by storage in leaves or transfer to roots for storage or metabolism (OH^- to soil)
	Oxalate production, storage in leaves or precipitation as Ca^{2+} salt
	Carbonate production and precipitation as salts
Time	Reduction on arrival in root
	Reduction after storage in roots, stems or leaves

regulation (shoot nitrate reduction). The energetics of the various possibilities in terms of photons used for the assimilation of nitrate and associated processes such as pH regulation have been discussed in detail by Raven (1985).

Although uptake and assimilation of ammonium are energetically much more favourable than for nitrate, many plants grow better using nitrate than ammonium. This is thought to be because ammonium in all but low concentrations is toxic. However, high concentrations are not usual in soils where there is exchange with colloids. Because nitrate assimilation can be varied in both space and time within a plant, it has a versatility which may be of advantage in certain environments (Table 1.13 and Sprent, 1987).

2

Evolutionary and current constraints on the nitrogen cycle

Evolutionary considerations

The early atmosphere is often considered to have been a reducing one, in which ammonia was the main form of nitrogen (Hart, 1979), although the stability of ammonia in the presence of high levels of ultraviolet radiation has been questioned (Canuto, Levine, Augustsson & Imhoff, 1982). This ammonia could have reacted with reduced carbon compounds such as methane, using energy supplied by electrical discharge or ultraviolet radiation to give simple nitrogen-containing molecules such as amino compounds and purines. From these compounds primitive organisms may have evolved. A very simple nitrogen cycle, as illustrated in Figure 2.1A, could have operated. Ammonia-assimilating systems would have been necessary, both to incorporate atmospheric ammonia into cells and also to recycle any ammonia formed following death of organisms.

If, as Levine & Augustsson (1985) have argued, the prebiological atmosphere consisted of N_2, CO_2 and H_2O, the situation would have been quite different. The implications of this hypothesis for nitrogen cycling will not be explored at this stage. One point which is generally agreed is that, early in biological times, nitrogen gas was present in the atmosphere. although probably at a lower concentration than now.

Denitrifying organisms as we know them today could not have been present unless there was oxidized nitrogen for them to metabolize, as there is no known biological route directly from NH_3 to N_2. If, as Towe (1985) argues, an ozone layer was necessary to protect primitive (aquatic) organisms from ultraviolet radiation, then some oxidized nitrogen was likely to have been formed by reactions similar to those which still occur (chapter 1). Further, since anaerobic organisms exist now when the atmosphere is oxic, then it is argued that they could have done so in earlier times. However, where this oxygen might have come from in primitive

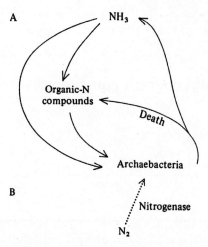

Figure 2.1. A, hypothetical primaeval nitrogen cycle; B, introduction of nitrogenase into archaebacteria.

times is not clear, so we shall adopt the more usually accepted idea that the early atmosphere lacked oxygen. Although nitrogen was available, biological nitrogen fixation in its present form was unlikely to have evolved at this stage. This is because the high energy requirement for the synthesis and operation of nitrogenase coupled with its slowness rendered it uncompetitive until nitrogen became the major limiting element. This occurred after the evolution of biological carbon dioxide fixation (Sprent & Raven, 1985). If, however, early photosynthetic organisms were anaerobic bacteria whose structure was similar to that of their extant relatives, they would have had a low C : N ratio, since, apart from storage material (e.g. glycogen, poly-β-hydroxybutyrate) and extracellular polysaccharide, bacterial cells are constructed largely from nitrogen-containing compounds (Table 2.1). Nitrogen fixation at this stage would have been a great advantage, since carbon compounds as a source of energy were not limiting and supplies of abiologically formed ammonia would have been rapidly used. Even if some abiologically generated oxidized nitrogen was available, this too could have been rapidly used. Thus we can put another step into the nitrogen cycle (Figure 2.1B). Once both atmospheric CO_2 *and* N_2 could be fixed, a very rapid increase in global biomass became feasible. However, shortly (in geological terms) after this, cyanobacteria, using oxygenic photosynthesis (H_2O as electron donor) are thought to have appeared on the scene. The evolutionary pressure towards this may have resulted from a shortage of reduced sulphur for bacterial photosynthesis (Towe, 1985). Many more possibilities and associated problems were then opened up and one major problem was that nitrogen-fixing organisms needed to keep oxygen away from nitrogenase (Table 1.4). When living organisms began to colonize land, the solution of this problem could be coupled to prevention of desiccation and of acquisition of trace elements and other nutrients. Nitrogen-fixing symbioses, with the fixing partner being protected and nourished by a eukaryote, became highly competitive.

As soon as oxygen gas was released into the atmosphere all the remaining steps of the complete nitrogen cycle became possible – not forgetting that abiological combination of nitrogen and oxygen was likely then, as now. Thus *Nitrosomonas* and *Nitrobacter*-like organisms would have been present. Recent evidence that these two genera are evolutionarily widely separated poses an interesting paradox here (Stackebrandt & Woese, 1984). During early colonization of land, a major net input of combined nitrogen into global biomass was needed. As biomass tended to stabilize at its present level, significant nitrogen fixation became necessary only in pioneering sites and those where much nitrogen is removed by cropping or fire, and to balance denitrification. Had plants not evolved structural material largely free of nitrogen (cellulose and other polysaccharides, lignin), the decline of nitrogen fixation might have been delayed. Perhaps it was the expense of nitrogen fixation that gave a selective advantage to autotrophs with structural elements not containing nitrogen. We arrive at the present situation where nitrogen gas is the most abundant form of nitrogen and yet fixation per unit of global biomass may be at a minimum. This alone indicates the advantages of nitrogen cycling.

Current constraints
The significance of C:N ratios
Carbon to nitrogen ratios vary widely from organism to organism, being lowest in animal tissue, then microbial (with the lower end similar to animals), then plant (Table 2.1). However, protoplasm itself, disregarding structural and storage material, has a rather constant C:N ratio, as might be expected. Any aerobic non-autotrophic cell or organism in acquiring 'food' is unlikely to find it with a C:N ratio exactly matching its needs. Excess or unusable nitrogen may be excreted and, because much of it is soluble, generally turns over rapidly in the nitrogen cycle. A slight excess of carbon is normally needed to balance that lost as respiratory carbon dioxide. A large excess of carbon-containing food can

Table 2.1. *C:N ratios (weight: weight), relative to 1, for various substances and organisms. A & S denotes Atkinson & Smith (1983); S & R denotes Sprent & Raven (1985)*

Compound/organism	C:N	Reference/comment
Amino acids, amides	1.7 to 7.7	Alanine to glutamine
Purines, pyrimidines	1.1 to 1.7	C_5N_4 to C_4N_2
Nucleic acids, typical	3	
Protein, typical	3.1	
Eubacteria	3.7 to 13.8	S & R, Nagata (1986). Higher values often associated with extracellular polysaccharide production
Cyanobacteria	7 to 26	A & S, S & R
Insects	4.8 to 5.7	S & R
Coelenterate medusae	3.5	Larson (1986)
Sheep's heart	3.6	S & R
Chlorophyta	7 to 24	A & S, Niell (1976)
Rhodophyta	6 to 11	A & S, Niell (1976)
Phaeophyta	9 to 55	A & S, Niell (1976). Larger values usually associated with macroalgae such as *Sargassum*
Leaves, living	5 to 21	S & R
Leaves, senescent	18 to 150	various
Oat caryopsis	30	S & R
Grain legume seed	10 to 15	S & R
Secondary woody tissues	50 to 1000	S & R
Lichens	12 to 23	S & R. Lower values often from N_2-fixing spp.

cause problems because it must be voided. Motile animals can walk/swim/fly away from the problem. Sedentary ones are largely ruled out of a terrestrial environment on this account. Plants, however, by using excess carbon for structural purposes can husband their nitrogen resources to make protoplasm and yet remain sedentary. However, use of carbon compounds for structural purposes involves production of complex polymers such as cellulose and lignin. The mechanical properties required of these polymers are met by substances which are not readily accessible to enzyme attack, a feature which in itself is advantageous to the plant. A comparison of the structures of cellulose and amylose (Figure 2.2), both 1,4-linked polymers of glucose, illustrates this point. Cellulose is much less easily digested than amylose (one of the two major components of starch), hence the need for cellulolytic microorganisms in guts of herbivores. In considering the operation of the nitrogen cycle, we should therefore not only take into account the C:N balance of the substrate but

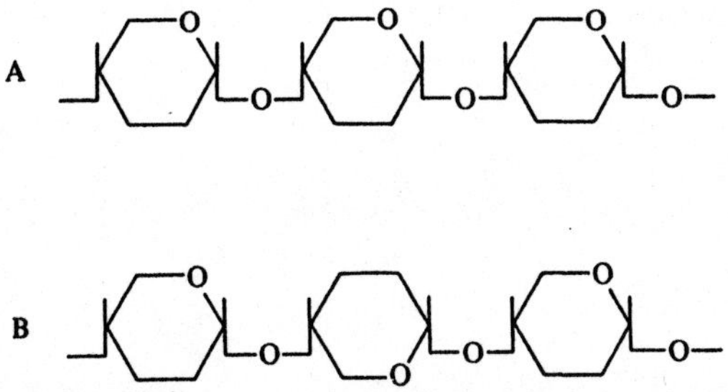

Figure 2.2. Diagram to show the difference between amylose (A) and cellulose (B). Both are composed of chains of glucose units, linked by their 1 and 4 carbon atoms. In amylose the linkage is an α one: the three-dimensional structure becomes a spiral. In cellulose the linkage is a β one which effectively means that alternate glucose residues are 'upside down'. This has the effect of cancelling out the spiral so that the molecule is essentially linear and can be closely packed with other similar molecules to form fibrils, which are resistant to enzyme attack.

also the structure of the carbon and nitrogen-containing components. Substrates where much of the carbon is structural rather than metabolic are less readily broken down. These arguments also hold when the structural component is a nitrogen-containing substance such as chitin, found in the exoskeleton of arthropods and in the cell walls of most fungi. Figure 2.3 illustrates the general cases of biomass whose metabolic material has a C:N ratio of 5 or 3 and a structural C:N of 200 or 50 respectively. The curves were derived using the formula of McGill, Hunt, Woodmansee & Reuss (1981):

$$\text{Fraction of N in structural component} = \frac{(C:N)_D - (C:N)_M}{(C:N)_S - (C:N)_M}$$

where $(C:N)_D$, $(C:N)_M$ and $(C:N)_S$ are the C:N ratios of dead, metabolic and structural components respectively. The fraction of carbon is obtained in a similar manner, using N:C ratios. Because carbon is greatly in excess of nitrogen in all tissues, the relationship for the fraction of carbon is essentially linear. This is not true for nitrogen, where a relatively small increase in %N in dead tissue reflects a major switch from structural to metabolic pools. This, as well as the nitrogen content *per se*, is why plant litter with a higher %N is more easily mineralized than that with lower %N. Similarly, we would expect microbes to be mineralized more readily than higher plants and, within the microbial components, bacteria than fungi. Further, because plants in aquatic environments have less, and less complex structural material than those in terrestrial environ-

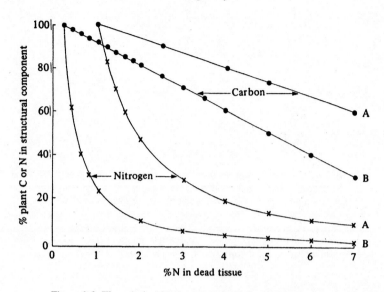

Figure 2.3. The relationship between the %C and %N in structural components and %N in dead plant tissue. Two cases are given: A, structural C:N of 50, metabolic C:N of 3; B, structural C:N of 200, metabolic C:N of 5. Values for carbon are plotted as solid circles; for nitrogen, plotted as crosses, the corresponding N:C values were used. The curves were derived using the formula of McGill *et al.* (1981).

ments, we might expect nitrogen cycling to be more rapid there. However, there are instances where plant residues with similar C:N ratios mineralize at very different rates. Figure 2.4 shows data for breakdown of various hardwood evergreen leaves in the Himalayas (Upadhyay & Singh, 1985). Rates of decay (indicated by a rise in %N and weight loss with time) were inversely related to the lignin content of the leaves. All of the above arguments assume that other constraints (e.g. pH or temperature extremes) are absent. In some systems, as we shall see later, these constraints may have an overriding influence.

Availability of co-factors

Many essential enzymes of the nitrogen cycle contain metals (Table 1.3). Whilst a requirement for iron (which was probably plentiful, especially in combination with sulphur, in early times (Towe, 1985)) was incorporated into numerous enzymes, molybdenum and nickel are especially important in the context of nitrogen metabolism. The fact that both nitrogenase and nitrate reductase contain molybdenum, suggests that this element was relatively freely available when these enzymes evolved.

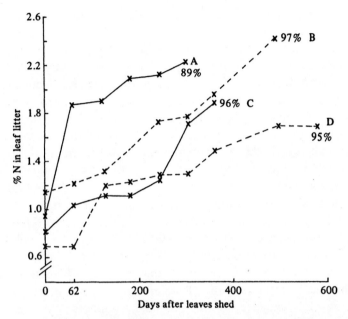

Figure 2.4. Effect of lignin content on rate of breakdown of leaves of four evergreen species (from Upadhyay & Singh, 1985). The percentage values given are the fresh weight losses after the appropriate time. The species and lignin content of leaves (% dry weight) are: A, *Quercus glauca* (10.8%); B, *Q leucotrichophora* (16.7%); C, *Lyonia ovalifolia* (15.8%); D, *Rhododendron arboretum* (17.9%).

In dry areas of the world, combined nitrogen is generally in the form of nitrate (Sprent, 1985). If none is present, nitrogen fixation may be an advantage: in either case molybdenum is required. However, nitrogen-fixing plants generally need more molybdenum than nitrate-assimilating plants of the same species. This is because of the slowness of nitrogenase compared with nitrate reductase. It takes approximately one hundred times as long per unit of molybdenum to transform a mole of nitrogen using nitrogenase as it does using nitrate reductase (Sprent & Raven, 1985). In some areas of the world, molybdenum is deficient. Thus legumes, cereals and other crops have been found to respond markedly to additions of molybdenum in Australia (Lipsett & Simpson, 1973), parts of the USA (Doerge, Bottomley & Gardner, 1985) and Brazil (Franco & Munns, 1981); incipient molybdenum deficiency has also been suggested for Nigeria (Lombin, 1985). Other parts of the world have levels of molybdenum which are potentially toxic to animals. It would be interesting to relate these geographical differences to the evolution of different forms of nitrogen nutrition in plants.

Although nickel is probably an essential element for all organisms, it has only recently been added to the list for plants, when it was shown to be a constituent of urease (Eskew, Welch & Cary, 1983). Cases of nickel deficiency have rarely been recorded. Nickel in excess can also be toxic and its effects on animals tend to be reciprocal with those of molybdenum – high molybdenum gives nickel deficiency symptoms and vice versa. Thus legumes, because of their often high concentrations of molybdenum, may harm herbivores. These interactions could have subtle effects on the nitrogen cycle, but these have yet to be explored.

Although many enzymes contain iron, nitrogenase has a particularly high content (about ten times that of nitrate reductase, for example): this, coupled with the slowness of nitrogenase, may mean that nitrogen-fixing organisms are iron deficient in marine environments (Rueter, 1982).

Availability of many elements varies with pH. For example, molybdenum is less available in low than high pH environments. Under some conditions, particularly when growing on ammonium or fixing nitrogen, plants acidify their substrates. Generally, plants growing on nitrate make their substrates less acid (see Raven (1985) and references therein, also chapter 3). These interactions between nitrogen nutrition and availability of trace elements may be important for nitrogen cycling in poorly buffered soils. Another factor which has recently been shown to affect molybdenum nutrition is sulphate. This may be particularly significant in marine waters where sulphate has been shown to inhibit the uptake of molybdenum by marine phytoplankton (Howarth & Cole, 1985). The effect may be sufficient to render many marine systems nitrogen limited (see also chapter 6), whereas their freshwater counterparts are more likely to be phosphorus limited, even though seas generally have higher concentrations of molybdenum than freshwater lakes or streams.

Co-evolution of organisms: symbiosis and syntrophy
Symbiosis
Within the context of the nitrogen cycle, the most important symbioses are those which fix nitrogen. Most of these are between nitrogen-fixing organisms and green plants and, with the exceptions of diatoms and *Azolla* (Table 1.9), are terrestrial. Although many nitrogen-fixing organisms can survive long periods of desiccation, metabolism, including nitrogen fixation, is reduced to a very low level. This desiccation tolerance is even put to use in the preparation of *Rhizobium* inoculants for use in agriculture. Plants, in colonizing dry land, which was coupled with the increasing ascendancy of the sporophyte generation, have developed

homoiohydry to a very marked degree (Raven, 1977). Thus a niche inside a plant provides a moist environment for nitrogen-fixing organisms. This, as well as the ready supply of carbon compounds, has led to the pre-dominance of symbiotic nitrogen fixing systems on land today (Sprent & Raven, 1985). Because land plants have developed extensive root systems, which may be modified in various ways in particularly low-nutrient soils (Lamont, 1982), to scavenge for both water and nutrients, symbiotic nitrogen-fixing microorganisms can also reap the benefit of a ready nutrient supply. In all cases the microorganism fixes nitrogen and the host plant carbon dioxide. In advanced symbioses (actinorhizas, legume nodules), at least, ammonium assimilation also occurs in the host. Animals on land generally void excess carbon in their diets. However, there are two well-authenticated examples of nitrogen-fixing symbioses, involving animals which have problems in this respect. These are arboreal termites, living within the confines of a nest situated on a tree trunk, and shipworms (chapter 1). The latter are particularly interesting in that the endosymbiont is intracellular, halophytic and cellulolytic. This last property is unique – other systems all require a soluble carbon source.

Looser associations

Many bacteria associate with roots and leaves of higher plants. These are not true symbioses, even though in the case of some root associations the bacteria enter the root cells (Okon & Kapulnik, 1985). As yet, the metabolic relationships between the organisms have not been elucidated. In particular, direct transfer of fixed nitrogen to the host, without death and breakdown of bacteria, has not been shown. Additionally, many of the bacteria (*Azotobacter, Azospirillum*) produce substances which may stimulate plant growth. Whether plants which are so stimulated can spare more carbon to support bacterial growth has not been conclusively demonstrated. Production of plant growth stimulants would be a good investment on the part of the bacterium if this were to be proven.

Loose associations of this type are unlikely to have been a stage in the evolution of nodule-like symbioses, since the microorganisms involved are unrelated (Sprent & Raven, 1985).

Syntrophy

This term is generally applied to associations between like or similar organisms, unlike symbiosis where the organisms are very different. Syntrophic associations have mutually complementary physiologic and metabolic systems. They can occur within populations of one organism, for example *Bradyrhizobium japonicum* (Ludwig, 1984). In this case, two metabolic types of cell may be found: one fixes nitrogen, exports ammonium, but does not grow, the other assimilates ammonium, grows and divides. In theory at least, such associations might fix nitrogen under natural conditions. This is not generally possible with only one metabolic type, since induction of nitrogenase in most rhizobia appears to be coupled with the repression of enzymes for ammonium assimilation, the latter process being carried out by the host plant.

There are a number of examples where cellulolytic and nitrogen-fixing organisms exist in mixed populations. These range from protozoa and nitrogen-fixing organisms in the gut of some termites, to fungi with nitrogen-fixing organisms in leaf litter. This latter type of association has potential for turning waste plant material into animal feed and fertilizer (Veal & Lynch, 1984). A further example of this type of complementation is found between *Arthrobacter* and *Corynebacterium* which together carry out more heterotrophic nitrification than can either organism on its own (Rho, 1986). No doubt many more of these associations exist in natural environments.

The role of small animals

Protozoa have been mentioned a number of times. They generally feed on bacteria and are thus an important component of the microbial biomass. Only recently have attempts been made to quantify their role in the nitrogen cycle. In one study, Clarholm (1985) examined (in all combinations) interactions between wheat roots, bacteria and protozoa on nitrogen mineralization and plant growth in a loamy topsoil. Table 2.2 gives data showing the effect of protozoa on nitrogen contained in wheat. In all cases plants contained most nitrogen when protozoa had been added to soil. Grazing by soil invertebrates has also been considered and modelled for acid forest soils, where it was found to contribute significantly to mineralization (Anderson, Leonard, Ineson & Huish, 1985; see Ingham *et al.* 1986a,b for a similar study on a semi-arid grassland). The type of animal involved in litter breakdown varies generally with latitude, tending towards large invertebrates (annelids, molluscs, arthropods) at lower latitudes. Many of these (earthworms, termites) may also take litter from the surface down into the soil; this may enhance the rate of mineralization, especially in dry areas. In the latter, desiccation-resistant invertebrates, such as termites, play a crucial role, comparable to that of earthworms in wetter areas (Freckman & Whitford,

Table 2.2. *Effect of protozoa on accumulation of nitrogen (mg) by wheat plants grown in substrate with added bacteria. Some treatments also had added carbon and/or nitrogen sources*

Protozoa	Treatment (addition) (mg)			
	None	14.4 mg C	3.6 mg N	14.6 mg C + 3.6 mg N
Absent	1.61	1.78	2.65	1.67
Present	2.55	3.04	4.13	3.74

1987). Foliar herbivores, such as cattle, may modify plant nitrogen distribution so that there are higher concentrations in roots; this in turn may encourage root herbivores such as nematodes (Seastedt, 1985). This shows that effects of animals on nitrogen cycling may be complex and not always obvious.

Some other biotic interactions involving steps in the nitrogen cycle
With nitrogen-fixing organisms

One example was given earlier, namely the effect of nitrogen-fixing organisms on plant growth via production of growth substances. In addition to this, microorganisms may interact in various ways to affect nodulation processes: Table 2.3 summarizes some of these ways. Various explanations are possible, the most likely being, as before, via growth substances. For example, inhibition of nodulation of clover by *Azospirillum* can be mimicked by auxins (Plazinski & Rolfe, 1985b). Alternatively, effects may be via mineral nutrition. This could be the case for

Table 2.3. *Interactions between microorganisms which affect nodulation processes*

System	Effect	Reference
Alnus, Frankia, Pseudomonas	Nodulation enhanced by *Pseudomonas*	Knowlton & Dawson (1983)
Legumes, *Rhizobium, Azotobacter*	Nodulation enhanced by *Azotobacter*	Burns, Bishop & Israel (1981)
Legumes, *Rhizobium, Azospirillum*	Nodulation enhanced by *Azospirillum*	Tilak, Singh & Rana (1981)
	Nodulation enhanced or reduced according to strain of *Azospirillum*	Plazinski & Rolfe (1985a,b)

pseudomonads and *Azotobacter* which are known to be good at scavenging soil iron because they produce large quantities of iron-complexing compounds known as siderophores. Neilands & Leong (1986) have reviewed siderophores in the context of plant nutrition. Other organisms, such as *Pseudomonas aeruginosa* (Stojkovski, Payne, Magee & Stanisich. 1986), bind molybdenum to extracellular slime and thus make it unavailable to plants. This could affect both symbiotic nitrogen fixation and nitrate assimilation.

Effects of higher plants

These may produce substances which inhibit reactions of the nitrogen cycle, although earlier suggestions that tannins in litter inhibit nitrification (Rice & Pancholy, 1974) have recently been discounted (McCarty & Bremner, 1986). Black walnut (*Juglans niger*) produces a phytotoxin, 5-hydroxy-1,4-naphthoquinone, commonly known as juglone. Amongst other species, juglone inhibits the growth of *Alnus glutinosa* more than *Eleagnus umbellata*. Both of these are nitrogen-fixing actinorhizal plants. For reasons which are not yet clear, juglone breaks down more easily under *Eleagnus* than under *Alnus* (Ponder & Tadros, 1985), illustrating the complexity of such effects.

Rate-limiting steps, pool sizes and turnover rates
Rate-limiting steps

Taken overall, any reaction in the nitrogen cycle may act as a rate-limiting step and hence control the overall process. Such limitation may stem from a number of sources, summarized in Table 2.4. In biochemical terms, pathways are very often controlled by negative feedback systems, i.e. the product of a sequence of reactions, when it

Table 2.4. *Factors which may affect rate-limiting steps in the nitrogen cycle*

Factor	Example
Environmental conditions unsuitable for growth of particular organisms	Excess or deficiency of O_2, pH, H_2O or heat may differentially affect N cycle organisms
Inhibition of one organism by another	Various allelopathic effects, e.g. natural nitrification inhibitors
Lack of substrate	Insufficient free NH_4^+ in some soils for nitrifying bacteria
Lack of essential co-factors	Mo deficiency on N_2 fixation and NO_3^- reduction

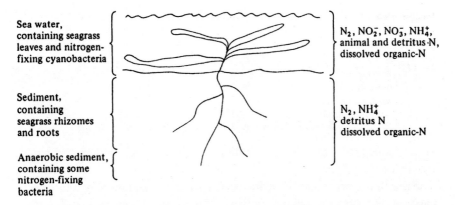

Sea water,
containing seagrass
leaves and nitrogen-
fixing cyanobacteria

N_2, NO_2^-, NO_3^-, NH_4^+,
animal and detritus·N,
dissolved organic-N

Sediment,
containing
seagrass rhizomes
and roots

N_2, NH_4^+
detritus N
dissolved organic-N

Anaerobic sediment,
containing some
nitrogen-fixing
bacteria

Figure 2.5. Features of a seagrass system used to model nitrogen cycling (Jones, 1985). Most of the nitrogen in the system was in the detritus fraction, but note that all other nitrogen-containing components used for plant growth are present. Reduction in these limits seagrass growth resulting in increased nitrogen fixation.

accumulates, slows or stops the sequence, usually by inhibiting an enzyme involved in an early step. In the nitrogen cycle, the step may be nitrogen fixation. A recent theoretical model argues that this is so for nitrogen-limited systems, and a seagrass community in Florida is used as an example (Jones, 1985). The system is essentially as illustrated in Figure 2.5. Seagrasses (monocotyledonous plants growing in marine coastal regions) have a high rate of biomass production, yet they contain only low concentrations of nitrogen (McComb, Cambridge, Kirkman & Kuo, 1981). The assumption made in the model, of nitrogen limitation, is thus probably justified. Although most of the nitrogen in the system passes through detritus, it is argued that when the system is perturbed, increased nitrogen fixation is the factor most likely to restore equilibrium. This is because it is the only stage which responds quickly and specifically to the levels of available nitrogen in the form of a negative feedback, i.e. when available nitrogen supply goes up, nitrogen fixation goes down and vice versa. Such a stabilizing role of nitrogen fixation could obtain for other nitrogen-limited systems.

One point of interest concerning the control of nitrogen fixation by negative feedback is that it is very unlikely to be controlled by substrate availability. Although nitrogen is not very soluble, it is not likely to be limiting in aquatic systems where the water is in equilibrium with an atmosphere containing 79% N_2. Even in the centre of the most active nitrogen-fixing nodules, N_2 is unlikely to be limiting (e.g. Sinclair &

Table 2.5. *Concentration of inorganic nitrogen in various waters of the UK (Anon., 1983). Ranges in mmol m^{-3}*

	NO_3^-	NO_2^-	NH_4^+
Estuaries	54–450	<1–6	6–143
Coastal waters	4–32	0–1	2–6
Some Scottish lochs	9–351	0–7	2–67

Goudriaan, 1981). Substrate limitation may well occur at other stages in the nitrogen cycle.

Overall, the processes of nitrification are aerobic and those of denitrification anaerobic. The balance between these is thus likely to be different in waterlogged and well-aerated environments. However, we should note that all but the very driest soils usually contain anaerobic microsites where denitrification may occur. The degree of oxygenation of aquatic environments depends upon many factors, including the presence of photo-synthetic (oxygen-evolving) organisms, the extent of stirring (wave action, turbulent flow in streams), populations of oxygen-consuming organisms. In most aquatic systems, nitrate is the predominant form of combined nitrogen (Table 2.5). In only two of the cases summarized was the concentration of ammonium greater than that of nitrate and these were for the Tyne estuary and the Teesside coast, both parts of north-east England where a considerable quantity of industrial effluent was being discharged into rivers. In global terms, 99% of inorganic nitrogen in oceans is in the form of nitrate (Rosswall, 1983).

Many of the microorganisms involved in the nitrogen cycle are either chemoautotrophs or photosynthetic. The major exceptions are some free-living and most symbiotic nitrogen-fixing species and some of the organisms involved in denitrification. Any of these may be limited by carbon substrates. Carbon limitations to nitrogen fixation have been discussed almost *ad nauseam* (see, for example, the review by Sprent & Minchin, 1985), even though many nodules are oxygen limited (Sprent *et al.*, 1987a). Evidence for the suggestion that carbon limits denitrification is less abundant, perhaps not surprisingly in view of the range of organisms involved (chapter 1). However, in the rhizosphere at least, denitrification may be related to plant photosynthesis in a way that varies with plant species. Beans (*Phaseolus vulgaris*) support more denitrification than maize (*Zea mais*) (Scaglia, Lensi & Chalamet, 1985), probably because legume exudates are richer in nitrogen than those of cereals (i.e. have a lower C:N ratio). The higher C:N ratio of cereal exudates may account

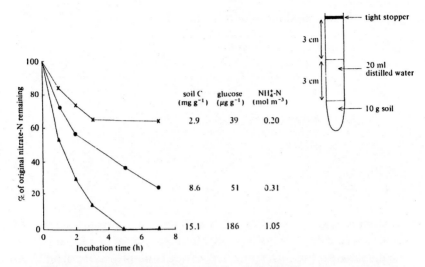

Figure 2.6. Effect of carbon content of soil on denitrification, measured as loss of added nitrate. Initial nitrate concentration was 6 mol m^{-3}. Soils were incubated in 50 ml centrifuge tubes set up as illustrated at 35 °C on a shaking water bath. Soil glucose was that extractable in 1 h at 100 °C. Ammonium nitrogen was estimated at 48 h. From Stanford, Vander Pol & Dzienia (1975).

for the large number of free-living or 'associative' nitrogen-fixing bacteria found around cereal and other grass roots.

Further evidence that carbon supply may limit denitrification in soils comes from work with added carbon sources, for example glucose or straw, or with soils varying in carbon content, but with excess nitrate added. Data of the latter type are given in Figure 2.6 (Stanford, Vander Pol & Dzienia, 1975). Denitrification was measured by loss of nitrate and was highly correlated with organic carbon in soil and even more highly with glucose (there are obviously some soils in which organic carbon is mainly humic and not readily available for microbial growth). There is a further correlation between ammonium production and organic carbon, and this raises an interesting question since the denitrification process as generally envisaged leads to nitrogen gas. It appears that dissimilatory nitrate reductase may be involved, not only with denitrification and nitrate respiration, but also in processes whereby ammonium is released into the environment – this would act as a nitrogen-conserving system. The conditions under which the normal denitrification pathway is switched to one producing ammonia are not yet defined (Stanford, Legg, Dzienia & Simpson, 1975, see also Knowles, 1982 and p. 25). In truly aquatic systems (rather than flooded soils), such as sediments at the

Table 2.6. *Content of two anaerobic sediments. From Kessel (1978). Components given as mg per g dry weight of sediment*

| | Sediment | |
Component/property	A	B
Organic matter	34.0	9.3
NH_4^+	0.14	0.03
Organic-N	1.63	0.17
pH	7.45	7.10

bottom of lakes, denitrification is also related to organic matter. Table 2.6. gives data from Kessel (1978). In the presence of added nitrate, the sediment with the lower organic matter content had its denitrification processes limited by carbon; the sediment with the higher organic matter eventually became limited by nitrogen. The rate-limiting step of nitrogen mineralization at pH values around 7 is the conversion of organic nitrogen to ammonium. The two subsequent steps are faster, therefore mineralization can and usually is measured by nitrate production (Smith, 1982).

Pool sizes and turnover rates
The estimated pools of nitrogen in global terms are given in Figure 2.7. Values are very approximate: they have been derived from many sources,

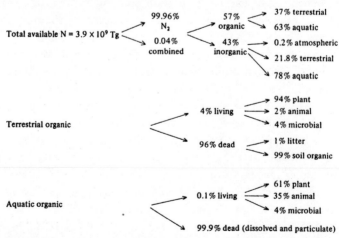

Figure 2.7. Approximate distribution of nitrogen in the world. Figures recalculated from Rosswall (1981, 1983).

as listed in Rosswall (1983). It is generally convenient to separate these into atmospheric, terrestrial and aquatic pools, and we shall adopt this procedure, before discussing their interrelationships. Although there is more nitrogen in the lithosphere (rocks, sediments, coal deposits) than all the other compartments put together (about 10^9 Tg), it is generally unavailable and will not be considered here. Small amounts may be released from coal as a result of burning; this will be discussed later.

Atmospheric pools

The atmosphere for the present purposes can be divided conveniently into two layers, the troposphere (which comprises 75% of the earth's atmosphere in terms of mass) and the stratosphere; the two are separated by the tropopause (Figure 2.8). The height of the tropopause varies greatly with latitude and to a lesser extent with season. On average the temperature decreases steadily with height in the troposphere (approximately 2 °C per 300 m), although local perturbations are common, depending on the prevailing meteorological conditions. Above the tropopause, temperature changes are much smaller and tend to show a slight increase with height. The tropopause thus acts as a rather stable boundary layer, and nitrogen cycle reactions in the stratosphere are often quite distinct from those in the troposphere. High levels of ozone (O_3) are present in the stratosphere and these affect many reactions involving nitrogen.

Overall, almost all (>99.9%) of the nitrogen in the atmosphere is in the

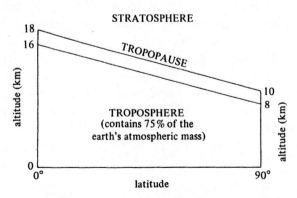

Figure 2.8. Diagram to illustrate the general structure of the earth's atmosphere. The boundaries are neither exactly linear, nor smooth. The exact height of the tropopause varies with season and weather; it is generally at its lowest in arctic winters.

Table 2.7. *Levels of inorganic nitrogen containing species which may be present in the atmosphere, with indications of sources, sinks and mean atmospheric lifetimes. From Hahn & Crutzen (1982) and Galbally & Roy (1983) All figures for nitrogen are in Tg. Time units: d, days; y, years.*

Species	Amount in atmosphere	Annual additions	Annual removals	Lifetime
$NO + NO_2$	0.2–1.0	23–80	23–80	1.5 d
N_2O	1400–1600	6.4–27.8	6.4–27.8	100–150 y
$HNO_3 + NO_3^-$	0.14–0.5	22–80	22–80	3–14 d
$NH_3 + NH_4^+$	0.8–3.5	55–80	55–80	7–14 d

form of nitrogen gas. However, this still leaves up to 1600 Tg of combined nitrogen, most of which is in the form of nitrous oxide (Table 2.7). Compared with other pools of combined nitrogen in soils and oceans, these amounts seem negligible, but this is definitely not the case. We must distinguish between pool size and turnover rate. A small pool size often means that the component is converted to something else very soon after it is formed – in other words its residence time or lifetime is short. In the case of N_2O, the mean atmospheric lifetime is thought to be of the order of a century. Apparently minor components such as other nitrogen oxides and ammonia have mean residences of only a few days. Nitric acid, for example, remains about 3 days: estimates for rate of formation are up to 80 Tg per year and it may be carried several thousand kilometres, mainly in a west to east direction, before being removed by rain. Hence arguments about who is responsible for acid rain with Norway, for example, putting much of the blame on the UK. Total amounts of substances which seem small thus may have far-reaching consequences. Similarly, the release of nitric oxide by supersonic aircraft (which, for operational reasons must fly at high altitudes, usually in the stratosphere) has given cause for concern over the stability of the ozone layer (Hahn & Crutzen, 1982). A further point to bear in mind here, as in aquatic and terrestrial reservoirs, is that distribution is not uniform. An obvious example is the exhaust stream of an aircraft, where concentrations of nitric oxide may be very high. Although diffusion is much more rapid in gas than liquid phases (10^4–10^5 times), it is not instantaneous and air masses can be extremely stable.

Terrestrial and aquatic pools
Approximately 37% of all organic and 22% of inorganic nitrogen is in the terrestrial compartment. Between this and the aquatic compartment

there are major differences in (a) the proportion of living: dead organic matter and (b) the proportions of plant: animal: microbial living biomass (Figure 2.7). Calculation of nitrogen in living biomass as a proportion of that in all the organic matter in the world suggests that about 1.5% is terrestrial and only about 0.6% oceanic. The microbial component as a proportion of the total living nitrogen is similar in aquatic and terrestrial systems. Plants are by far the major component, which is why they feature so prominently in this book, but animals figure more prominently in aquatic than terrestrial systems. In both there are major local differences. For example, organic matter may represent over 90% of soil dry weight in some peats and considerably less than 1% in many sandy and tropical soils. Turnover rates are probably equally varied.

Turnover rates
Estimates of turnover rates for different pools are difficult to make for technical reasons, such as lack of suitable isotopes and the complexity of systems where all parts of the cycle occur simultaneously. Recent advances in mass spectrometry have made field measurements with low levels of the stable isotope ^{15}N (using either ^{15}N additions or natural variations) more realistic. Combination of such measurements with mathematical modelling suggests that valid estimates of at least some nitrogen cycle reaction rates are possible (Myrold & Tiedje, 1986). Where one reaction has a much lower rate than the others (in this work,

Table 2.8. *Major movements of nitrogen species between atmosphere (A), terrestrial (T), freshwater (W_f) and marine (W_m) pools. W indicates* $W_f + W_m$

Process	Movement
Dry deposition	$A \rightarrow T$
Wet deposition	
Glacial deposition	$A \rightarrow W$
Leaching and surface run-off	$T \rightarrow W$
Gas transfer (lateral and vertical movement of air masses)	$A \rightarrow A$
Oceanic currents	$W_m \rightarrow W_m$
Human activities (fertilizer)	$T \rightarrow T$
Dust storms	$T \rightarrow T$
	$T \rightarrow W$
	$W \rightarrow A$
Volatilization/solubilization (NH_3)	$A \rightarrow W$
	$A \rightarrow T$

which was based on soil systems, denitrification was 10^{-3} times that of other reactions) it is likely to be estimated much less accurately. Specific examples of turnover rates will be given in later chapters.

Global transfers

Obviously nitrogen moves between atmospheric, terrestrial and aquatic pools. This section will consider large-scale movement of individual species of nitrogen compound within and between pools. Small-scale movements will be considered as special cases in Part II.

Table 2.8 summarizes the major features of global nitrogen transfer (see also Reiners, 1983). Variability is great. At the present time, glacial effects and oceanic currents are fairly constant. However, dust storms and anything related to human activity are very variable. Dust storms (many of which contribute to the enlargement of desert areas) show marked patterns of seasonal variation. Leaching and run-off are increasing steadily with the use of nitrogenous fertilizers. It is estimated that 11 Tg of dissolved inorganic nitrogen may pass yearly from rivers to oceans (Richey, 1983). Oceans are slowly but surely increasing in nitrogen content. The effects of this on long-term productivity remain to be assessed. One indication of the effects of man can be gauged from analyses of the upper and lower stretches of the river Rhine (Table 2.9). The river Scheldt, which flows into the North Sea near Antwerp in Belgium had 650 mmol m^{-3} of ammonium nitrogen, over four times the concentration of nitrate. As much as 30% of all added nitrogen fertilizer may eventually end in the ocean and further amounts pass to inland lakes. In contrast, the major tropical rivers such as the Amazon, Niger and Zaire, have 8 mmol m^{-3} of nitrogen, most of it as nitrate, but see also p. 95.

Table 2.9. *Inorganic nitrogen in the upper and lower reaches of the Rhine. From Richey (1983). Components given in* mmol m^{-3}

	Component	
	($NO_3^- + NO_2^-$)	NH_4^+
Sampling area		
Upper alpine	30	2.5
Lower, just upstream of Nijmegen	230	90

3

Shortened and open open nitrogen´cycles:
effects of environment

There are many examples in the present world where stages of the nitrogen cycle may be virtually omitted *at a particular site*. In addition, large areas of land or water may have open systems where nitrogen is brought in from one source and lost to another. Possible variations on this scheme are indicated in Figure 3.1. This chapter discusses ways in which such situations may come about. In some cases there is considerable overlap. The effects of the major environmental variables are also considered.

Open and shortened cycles

The importance of C:N ratios in determining the rate and operation of the nitrogen cycle has already been discussed. Most studies have concentrated upon the problems of mineralization when C:N is high. However, if we regard mineralization as primarily driven by the need of microorganisms for reduced carbon, then at very low C:N this reduced carbon can be very limiting and may greatly bias the nitrogen cycle. An extreme case is illustrated by the nitrogen cycle in the vicinity of a penguin rookery on Marion Island (the larger of the Prince Edward Islands, situated approximately half-way between the Cape of Good Hope and the mainland of Antarctica, 47°S, 38°E; Lindeboom, 1984). Here, when penguins come ashore to moult and lay eggs, they excrete large quantities of uric acid (which has a C:N ratio of 1:1.25). In obtaining energy from uric acid, microorganisms release large quantities of ammonium. By the combined effects of high pH and strong winds, most of this volatilizes as ammonia. Some returns to the island in rain, and this supports the dense vegetation found around the penguin rookeries, but most is carried out to sea. Indeed, early sailors are said to have navigated the last 10 km to the island by the smell of ammonia (Figure 3.2). This ammonia could be washed into the sea and be available for

A   Steady state

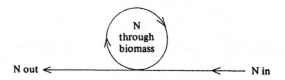

B   Some N immobilized

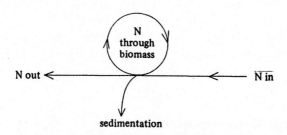

Figure 3.1. Diagram to illustrate open nitrogen cycles. For each, the incoming and outgoing nitrogen may or may not be in the same form (compound).

marine organisms. Whether it ever passes through the food chain back to the penguins is not known. If not, we have an open 'cycle' where nitrogen is brought into the system as penguin and leaves it as ammonia. A further interesting feature of this system is the final fate of the ammonia nitrogen which is returned to the island and used by vegetation. Since there is no significant herbivore component present, microbial breakdown of dead vegetation is very low. This results in a build up of peat, which effectively immobilizes the nitrogen.

A different, less extreme type of open 'cycle' occurs in flowing streams. Where these occur in forests, the amounts of nitrogen involved may be considerable. A nitrogen budget for a watershed stream of a conifer forest in Oregon, USA, was drawn up by Triska *et al.* (1984). Nitrogen inputs to the system were from subsurface flow, mainly organic (10.6 g m^{-2}) but with some nitrate (0.5 g m^{-2}). Litterfall, lateral movement and some nitrogen fixation associated with fire-wood debris resulted in a total input

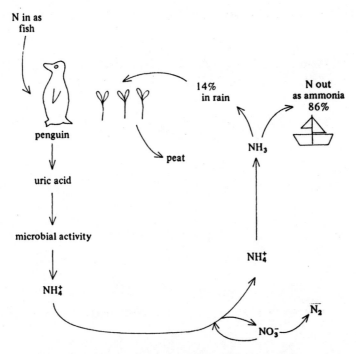

Figure 3.2. Generalized diagram of the open situation obtaining on Marion Island, Antarctica. Two species of penguin are found, *Aptenodytes patagonius* and *Eudyptes chrysolophus*. Details of cycling vary between species and can be found in Lindeboom (1984). Some nitrogen also leaves the island as young penguin.

of 15.3 g m^{-2} y^{-1}. Total annual output was estimated to be 11.4 g m^{-2} so there was a net nitrogen gain. Within the system all nitrogen cycle reactions appeared to operate. Unlike the system described above, C:N ratios were very high due to the wood component. Both systems have marked temporal variations, in the present case input of debris by storms, which occasionally (at intervals *c.* 500 y) result in debris torrents each estimated to be equivalent to 30% of that of the intervening 499 y.

Omission of specific steps

In many natural climax situations including forests and prairie grasslands, one or more steps in the nitrogen cycle may be omitted. Such systems usually have very tight nitrogen cycling, resulting in little or no loss, therefore no need for nitrogen fixation. Frequently this tight cycling is achieved by cutting out the nitrification step and thus bypassing nitrate,

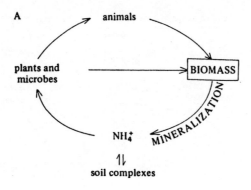

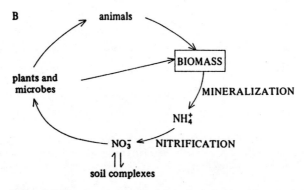

Figure 3.3. Shortened nitrogen cycles. A, omission of the nitrification steps: no net gains by fixation or losses by denitrification or volatilization occur. B, when soil nitrate is complexed: again there are no net gains or losses.

the most readily leached component. The principal form of mineral nitrogen is ammonium, much of which is complexed on soil colloids (Figure 3.3A). Bypassing the nitrification step may be achieved by some of the plants producing nitrification inhibitors or by mycorrhizas (see below).

Although in most soils ammonium is the least mobile form of nitrogen, occasionally nitrate is complexed (Figure 3.3B). This occurs in some acid soils of Central and South America (Kinjo & Pratt, 1971). The amount of nitrate adsorbed is inversely related to soil pH, and nitrate can be displaced by other anions such as sulphate and phosphate. The reason for nitrate adsorption appears to be the presence of high levels of amorphous inorganic materials such as SiO_3 and Al_2O_3. Up to 8.7 milliequivalents of nitrate per 100 g dry soil may be complexed. Whether nitrogen fixation has a role in such systems will depend on whether or not the biomass is aggrading and whether there is any nitrogen loss. It is interesting to note that the high pH values of these soils would tend to encourage ammonia volatilization, so perhaps it is as well that soil nitrogen is complexed following nitrification, rather than nitrification being inhibited.

Mycorrhizas

Mycorrhizas may effectively help to bypass reactions of both ammonification and nitrification. There is a considerable quantity of evidence that they can take up ammonium and some that they can take up simple organic nitrogen (Bowen & Smith, 1981; Alexander, 1983). In the case of ammonium, if the mycorrhizal fungus competes successfully with other soil microorganisms, the net effect will be the same as when plants produce nitrification inhibitors – omission of the nitrification step. If organic nitrogen is taken up directly by mycorrhizas, it will not be available for ammonification. Further, since mycorrhizal hyphae may form a continuum, connecting different host plants, the possibility of direct transfer of nutrients between plants exists (Francis & Read, 1984).

Interaction with other minerals

We have seen in chapter 2 that molybdenum may be unavailable in saline waters because of high levels of sulphate and that this may limit nitrogen fixation. Under some conditions soil minerals may actually inhibit reactions of the nitrogen cycle. In a study of the French Hautes-Vosges, Boudot & Chone (1985) found high rates of mineralization and nitrification (as measured by content of mineral nitrogen) in one soil and low rates in another, although both showed similar carbon mineralization rates (measured by carbon dioxide evolution). The reasons for this were investigated in a series of experiments which included the addition of $(^{15}NH_4)_2SO_4$ and $Na^{15}NO_2$. The results (Table 3.1) show that the andic soil can generate nitrate from ammonium and is also better able to generate nitrate from nitrite than the colluvial soil. Nitrogen budgets suggest that much of the nitrite is complexed with organic matter, and that some undergoes chemodenitrification. Thus, although some of the reactions of mineralization/nitrification occur, inorganic nitrogen bound to organic matter (in the form of ammonium and nitrite) is unavailable for biological processes. The $C:N$ ratio of the organic matter of such soils is higher than would be expected for soils in which mineralization is absent. The complexing of mineral nitrogen is thought to result from the high levels of aluminium in the more weathered soil. It is thus clear that

Table 3.1. *Possible effects of amorphous aluminium in soil on mineralization and nitrification. Data from Boudot & Chone (1985)*

	Soil type	
	Andic	Colluvial
Underlying rock	Graywackes	Granite
pH (aqueous extract)	4.5	4.5
% organic matter	46	14
% $Al(OH)_3$ extractable in oxalate buffer	1.0	0.4
% ^{15}N recovered from added $(^{15}NH_4)_2SO_4$ after 42 days		
as NH_4^+	58	52
as NO_3^-	1	20
organic	34	34
% ^{15}N recovered from added $^{15}NO_2^-$ after 65 days		
as NH_4^+	5	5
as NO_3^-	24	46
organic	56	49
% lost, possibly by denitrification	15	0

aluminium is a very important element in its interactions with the nitrogen cycle. In addition to those mentioned, aluminium is toxic to most rhizobia and hence inhibits nodulation and nitrogen fixation (see, for example, Whelan & Alexander, 1980), especially at low soil pH.

Obviously there will be interactions between the nitrogen cycle and cycling of other major and minor nutrients. Interactions between the N, C, P and S cycles on a global scale have been discussed in Bolin & Cook (1983). The cycling of these elements through decaying organic matter (soils and sediments) can be divided into two parts, according to the type of chemical bond to carbon, covalent or ester (Table 3.2). Covalent bonds are broken down by extracellular enzymes when microorganisms require energy. Thus mineral nitrogen and sulphur are produced, in a sense, as by-products, although when these elements are in short supply they are incorporated into the microbial biomass. Both sulphur and phosphorus occur in ester bonds (phosphorus almost exclusively so). These bonds are hydrolysed, again by plant and microbial extracellular enzymes, releasing sulphate and phosphate which are utilized by the organisms. This means that in some environments soluble inorganic sulphur and phosphorus compounds are virtually absent (although rock phosphate may be present).

Phosphorus has been studied mainly with respect to nitrogen fixation,

Table 3.2. *Bonds in phosphorus, nitrogen and sulphur compounds in decaying organic matter and their relation to mineral cycling*

Type of bond	Consequences for cycling
Covalent	
C–N, C–S, (P–C)	Broken when organisms use compounds as C substrates. Inorganic N and S released into soil. (Processes involving P less significant.)
Ester	
C–O–S, C–O–P	Broken when organisms need S or P

which because of its energy utilization is alleged to have a large requirement for phosphate. This prediction is borne out by the rapid response of clover in some pastures to phosphate (e.g. in the volcanic soils of New Zealand) and by rapid eutrophication by nitrogen-fixing cyanobacteria in some phosphate-rich waters (Stewart & Alexander, 1971). It is further claimed to be the basis for positive interactions between vesicular–arbuscular (VA) mycorrhizas and nodulated plants (e.g. Daft & El-Ghiami, 1976). VA mycorrhizas extend the search area for phosphate and also may solubilize rock phosphate. However, detailed analyses of responses suggest that mycorrhizas stimulate plant growth *prior* to phosphate uptake (Robson, 1983). This indicates that phosphorus is used for plant growth rather than nitrogen fixation *per se*. Further, analyses of plant material from different sources show that many legumes, in particular some tropical genera such as *Stylosanthes* (Probert & Williams, 1985), can grow and fix nitrogen with much lower levels of phosphorus per unit dry weight than others. Indeed, some plants will not grow at levels of phosphate considered to be necessary for agricultural crops (Reuter & Robinson, 1986). The ratio N:P may be very important in breakdown of complex nitrogen compounds in forest ecosystems (Vogt, Grier & Vogt, 1986).

Sulphur is an essential nutrient for all organisms, but sulphide can have major effects on denitrification, by inhibiting the final reaction to nitrogen gas (Figure 3.4).

In a recent review, Bloom, Chaplin & Mooney (1985) applied economic theory to resource acquisition and concluded that plants adjust their phenology to acquire resources when they are cheap, storing them if necessary until conditions are optimum for growth, and that generally plants avoid extreme ratios of essential elements (e.g. S:P). The argument that plants make tissues or organs in order to acquire the most

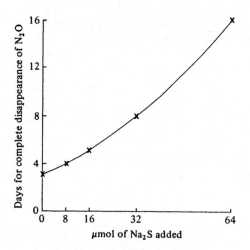

Figure 3.4. Effect of sodium sulphide on reduction of nitrous oxide by soil. Soil (10 g) was placed in a 50 ml Erlenmeyer flask, preincubated anaerobically for 2 days, then given 10 mg glucose and the indicated amount of Na_2S, then again incubated anaerobically (under helium) after addition of 20 μmol N_2O. From Tam & Knowles (1979). Note that the higher levels of sulphide are unlikely in most natural ecosystems.

limiting nutrients and hence restore overall balance is consistent with the production of nodules on legumes only when nitrogen is the major limiting nutrient (Sprent, 1985). When nutrients generally are limiting, resources are better used for production of extensive root and root hair systems, possibly together with specialized structures such as proteoid roots and mycorrhizas (Lamont, 1982; Robinson & Rorison, 1983). An alternative is to have a conservation system where nutrients may be retained within the plant, such as evergreen leaves or underground storage organs. The latter are found extensively in certain geophytes which have large underground nutrient (and often water) storing organs (Pate & Dixon, 1981). Under such nutrient-limiting conditions when there is a high degree of conservation, nitrogen cycling may be slow (low levels of substrate) and/or closely linked to herbivores which eat living plant material, rather than detritus-feeding animals.

Environmental effects
Water relations

The nitrogen cycle on land is very closely linked with water availability. Apart from its obvious role as a solvent, water deficiency affects components of the nitrogen cycle in different ways. Nitrification is

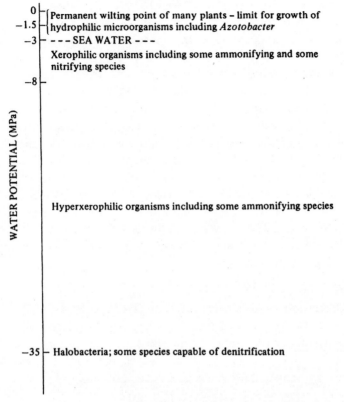

Figure 3.5. Lowest water potentials at which some organisms of the nitrogen cycle are found.

more sensitive to water stress than mineralization and this may result in transient increases in levels of ammonium during the early stages of soil drying (Dommergues, Garcia & Ganry, 1980). This sensitivity may relate to the high energy requirement of autotrophic nitrifying organisms diverting energy resources which might otherwise have been used to synthesize compatible solutes such as amines or polyols, which could help the organisms to withstand dry conditions. However, in due course the predominant form of nitrogen in dry soils is nitrate. Plants generally are most water use efficient when grown on ammonium (Raven, 1985), nitrate and N_2 following in that order. These facts are closely related to the energy required to acquire and assimilate these substances. However, as mentioned on p. 28, nitrate assimilation is a very versatile process and this versatility may offset its expense. Significant nitrogen fixation does not occur under very dry conditions (Sprent, 1985), but protection within a host plant may enable symbiotic nitrogen fixing microorganisms to survive until stress is relieved. The drought tolerance is then limited by the host, not the microorganism (Sprent, 1987).

The effects of water supply are difficult to separate from other environmental factors. In particular, high temperature is often accompanied by drought, and waterlogging in soils by anaerobiosis. Salinity may be a further complicating factor at all levels of water supply. Specific cases will be discussed in later chapters. However, some generalizations may be possible, for example that actinomycetes are more resistant to desiccation than aerobic bacteria (Smith, 1982b) and are thus more common in desert soils (Skujiņš, 1984). Wetting/drying cycles often have more marked effects than a continuous period at one water level for various reasons including (a) selective death of some microorganisms on drying (Figure 3.5) and (b) release of substances such as amino complexes from humic acids on drying (Seneviratne & Wild, 1985).

pH

All reactions of the nitrogen cycle involve pH changes, since they involve species carrying from 1 negative charge per N (nitrate, nitrite) through no net charge (N_2, urea) to one with a net positive charge per N (ammonium) (Table 3.3). In the latter case charge itself is markedly affected by environmental pH because of the equilibrium

$$NH_3 + H_2O \rightleftharpoons NH_4OH \rightleftharpoons NH_4^+ + OH^-$$

Thus the nitrogen cycle itself may affect the pH of the environment. A full discussion of these points can be found in Kennedy (1986). The degree to which environmental pH is affected by reactions of the nitrogen cycle depends on buffering capacity and the possible precipitation of some components, for example calcium carbonate. The situation is further complicated by the fact that some abiological reactions are favoured by pH extremes. In addition to all these problems, the enzymic reactions of the nitrogen cycle respond to pH.

For overall denitrification, the optimum pH is 7–8, but the process may proceed up to pH 11. At a pH of 3.5, denitrification is completely suppressed. At acid pH values above this the process stops at N_2O, rather than proceeding to N_2 gas. In other words pH affects the product of denitrification more than the process itself (Fillery, 1983).

Nitrification also has a pH optimum of 7–8. *Nitrobacter* is very sensitive to pH, largely because free ammonia and nitrous acid (the pK_a for $HNO_2 \rightleftharpoons NO_2^- + H^+$ is 3.4) are both very toxic to it. *Nitrosomonas* is

Table 3.3. *pH implications of nitrogen cycle reactions*
A, Nitrogen assimilation into microbial or plant cells
where the major nitrogen-containing component
(protein) carries a slight net negative charge (<1 per N)
at physiological pH values (see Raven & Smith, 1976)[a]

Process	H^+ or OH^- per N assimilated
$N_2 \rightarrow$ cells	$<1\,H^+$
$NO_3^- \rightarrow$ cells	$<1\,OH^-$
$NH_4^+ \rightarrow$ cells	$>1\,H^+$

B, Processes in soil or water, reactions not involving assimilation of nitrogen

Process	H^+ or OH^- per N processed
$NO_3^- \rightarrow N_2$	$1\,OH^-$
$NO_3^- \rightarrow NH_4^+$	$2\,OH^-$
$NH_4^+ \rightarrow NO_3^-$	$2\,H^+$

[a]Note that these are *overall* values. Local changes may be different. For example NH_3 is the initial product of N_2 fixation: this is immediately protonated, so that OH^- is generated at this site. (Kennedy, 1986)

slightly less sensitive to high pH or ammonia. Thus we have a variety of responses of nitrification to pH; these will depend also on the concentrations of free ammonia and of nitrous acid. These responses are summarized in Figure 3.6.

pH also has a marked effect on the volatilization of ammonia: increased pH increases volatilization, and the effects are more pronounced at high temperatures and wind velocities. Volatilization is also affected by buffering capacity (higher capacity, higher loss) and cation exchange capacity (higher capacity, lower loss). These factors are discussed in detail in the volume edited by Freney & Simpson (1983).

Nitrogen fixation is affected by pH in a number of ways. For example, legumes relying on fixed nitrogen are generally more sensitive to extremes of pH than are plants of the same species given mineral nitrogen, although there are exceptions (Lindstrom, Sarsa, Polkunen & Kansanen, 1985). Further, extremes of pH affect the survival of rhizobia in the soil. Free-living nitrogen-fixing organisms tend to be even more pH sensitive. This can severely limit nitrogen fixation in carbon-rich substrates such as forest litter. For example, Nohrstedt (1985) found that 85% of the

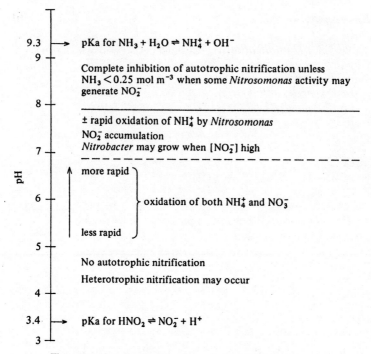

Figure 3.6. Summary of the effects of pH on nitrification.

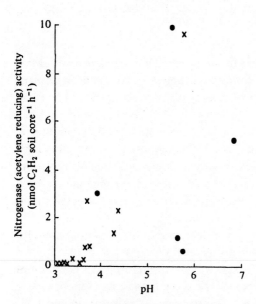

Figure 3.7. Effect of pH on nitrogenase activity in forest litter. Activity estimated on a 24 h incubation at soil temperature of cores 3 cm diameter and 3 cm deep from litter layer. pH was measured on 10 mol m^{-3} CaCl$_2$ extracts. Crosses = conifer litter (15 sites, 2 pairs of coincident readings); filled circles = broad-leaf litter. Data from Nohrstedt (1985).

variation in the yearly mean of nitrogen-fixing capacity from 15 stands of coniferous trees could be attributed to pH (Figure 3.7). Jones & Bangs (1985) showed that liming oak forest soils can markedly stimulate activity of heterotrophic nitrogen-fixing organisms such as *Clostridium butyricum*.

Aeration

This is clearly going to play a major role since denitrification is essentially anaerobic and nitrification aerobic. Since oxygen is used in preference to nitrate as a terminal electron acceptor by denitrifying and nitrate-respiring bacteria, low pO_2 would be expected to be a requirement for denitrification (this has a parallel in nitrogen-fixing organisms which will use combined nitrogen rather than fix their own).

Denitrification processes vary in their response to aeration, largely because of variations in the redox potential (E_h) of the different reactions. In general, dissimilatory nitrate reductases tolerate or even prefer low levels of oxygen. Synthesis of the enzymes catalysing reactions after nitrate reduction is derepressed by progressively lower oxygen concentrations. The E_h below which nitrate respiration begins is about 400 mV. Rates of denitrification may increase by 23% for every 100 mV decrease in E_h (Ashgar & Kanehiro, 1976). Under conditions where cell densities and organic matter are both high (e,g. in sewage sludge) denitrification may occur even at quite high *average* oxygen concentrations, because high rates of oxygen utilization and small anaerobic flocs give microsites with very low pO_2 (Focht & Verstraete, 1977).

Because the effects of aeration are largely *via* E_h, they would be expected to interact with pH: this is found to be the case (Focht, 1974). Further, in soils, the gas pore space volume affects the rate at which anaerobiosis sets in. Pore space in turn is greatly affected by irrigation and rainfall. Interactions between these factors mean that the amount of N_2O produced, relative to N_2 may vary greatly with changing environment (Focht, 1974). Decreasing aeration, which increases overall denitrification, reduces the $N_2O:N_2$ ratio because the rate of removal of N_2O exceeds its rate of production.

Heterotrophic nitrifying bacteria appear to function at lower oxygen concentrations than autotrophic species. The latter have an unusually high (for bacteria) $K_{1/2}$ for oxygen in respiration of 150–500 mmol m^{-3} (values of 1 are more typical). Thus at low pO_2 and especially where high organic matter levels obtain, heterotrophic may exceed autotrophic nitrification.

Aeration has major effects on nitrogen fixation, for reasons which have been discussed elsewhere (chapter 1). In summary, anaerobic conditions favour certain free-living species, both autotrophs and heterotrophs, whereas aerobic conditions are favoured by only a few free-living heterotrophs (e.g. *Azotobacter*, heterocystous cyanobacteria). Legume nodules may be oxygen limited as a result of internal diffusion barriers, even in well-aerated soils; the situation may be aggravated in dry, as well as wet soils (see Sprent *et al.*, 1987a, and references therein).

Salinity

Obviously organisms involved in marine nitrogen cycles must be tolerant of salinity. However, as we saw in the previous chapter, sulphate may limit molybdenum uptake and hence nitrogen fixation in marine habitats. Recent confirmation of denitrification by species of *Halobacterium* (Mancinelli & Hochstein, 1986) is consistent with the oceans being the major source of nitrogen for the atmosphere (see chapter 6). With

Table 3.4. *Effects of NaCl salinity on nitrogen cycle processes in soils. A, based on McClung & Frankenberger (1985) and references therein; B, based on Shipton & Burggraaf (1982) and Zahran & Sprent (1986)*

Component of N cycle	General effect	Comments
A Nitrification of either urea or $(NH_4)_2SO_4$	Considerable inhibition	Effect due to Cl^- since shown also by $CaCl_2$ but not Na_2SO_4. Amount of inhibition varies with soil, most on clay-loam
Ammonification by urea	Little or no effect	
Volatilization of NH_3 from $(NH_4)_2SO_4$	Little to major effect	Varies with soil: particularly high on soil with pH of 8.0
Volatilization of NH_3 from urea	No to little effect	Varies with pH and soil cation exchange capacity. Effect reduced by Ca^{2+} salts
B Symbiotic N_2 fixation	More inhibitory to host than either *Rhizobium* or *Frankia*	General effects greater on infection processes than on established nodules. Effect may be reduced by Ca^{2+}
C Denitrification	No data found on soils *per se*, but in view of the large number of denitrifying spp., unlikely to be a major problem	

increased irrigation, even using fresh water, salt concentrations in many soils are steadily increasing, so it is pertinent to consider salinity in a terrestrial as well as an aquatic context. The salts involved may be sodium chloride with or without salts of magnesium and calcium. This, and the differing nature of the soils themselves, means that there will be a spectrum of effects of salinity on nitrogen cycle reactions in soils. Table 3.4 summarizes some of these.

Temperature

Temperature will affect both the biological and abiological processes in the nitrogen cycle. Are the various processes differentially affected so that the cycle is biased differently under different temperature regimes? As with other environmental variables, temperature effects are likely to relate to the number of different species which can carry out particular reactions. Nitrogen-fixing organisms can be found in areas ranging from arctic regions to hot springs. However, in medium to high latitudes there are marked seasonal patterns of nitrogen-fixing activity which can be related to temperature as well as to water supply and photon flux density. Most nitrogen-fixing systems have a broad temperature optimum (spanning 10–15 °C), the actual range of which varies with the natural environment of the organism involved. However, as it interacts with so many other processes (e.g. respiration), fixation may not be equally efficient (in terms of carbon used per nitrogen fixed) over the whole temperature optimum range (see discussion in Sprent, 1979).

As far as denitrification is concerned, temperature seems to have a greater effect in soils than aquatic systems (Knowles, 1982), possibly because temperature fluctuations in soils are the greater. Rapid rises in rate of denitrification up to 25 °C are generally observed, followed by slower rises up to about 65 °C. The retention of activity at high temperatures has important management implications for waste treatment. Like pH, temperature may affect the final product formed (Figure 3.8). Longer incubation times increase the proportion of product as N_2 at low temperature over that shown in the figure, but even after 8 days at 28 °C, nearly 40% of gas evolved was N_2O. However, others have found little effect of temperature on relative quantities of product (see discussion in Focht & Verstraete, 1977). These apparent conflicts may result from different oxygen levels, incubation times, and type and pH of substrate. Both solubility and diffusivity change with temperature; these changes, which are particularly pronounced for gases, will have indirect effects on denitrification. Further, during prolonged incubation at higher temperatures, microbial populations will change in favour of thermophiles.

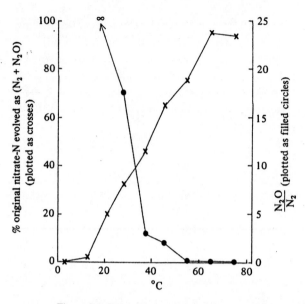

Figure 3.8. Effect of temperature on denitrification. The data are recalculated from table 10.1 of Payne (1981) and originate in measurements made in Sweden on a Vařing soil. For details and discussion see Payne (1981) and references therein. Soils were incubated for 2 days.

Perhaps not surprisingly, since only two principal genera are involved, autotrophic nitrification is less temperature versatile than either nitrogen fixation or denitrification. The organisms do not grow above 40 °C, but appear to be much more adaptable to low temperatures. Mineralization and nitrification occur particular rapidly when aquatic systems warm in spring and during freezing/thawing cycles in soils: the effects of freezing and thawing appear to result from a mixture of biological and abiological processes. There is a considerable body of evidence of strain variation in *Nitrobacter* and *Nitrosomonas*, with genotypes from cooler environments being more active at low temperatures than strains from warmer areas (Anderson, Boswell & Harrison, 1971). In many conditions the two genera respond similarly to temperature, but under some, *Nitrosomonas* is more sensitive than *Nitrobacter*: in this case, nitrite does not accumulate. However, as discussed earlier (see also Figure 3.6) these two genera vary in response to pH. Thus significant interactions between pH and temperature have been found and in some cases these do lead to an accumulation of nitrite (i.e. *Nitrobacter* is more affected than *Nitrosomonas*). Table 3.5 gives data which underline the effects of pH (Wong-Chong & Loehr,, 1975; see also discussion in Focht & Verstraete,

Table 3.5. *Effects of pH on activation energies ($E_a(kJ\,mol^{-1})$) for ammonium oxidation by* Nitrosomonas *and nitrite oxidation by* Nitrobacter. *From Wong-Chong & Loehr (1975)*

pH	E_a, NH_4^+ oxidation	E_a, NO_2^- oxidation
6.0	130.7	
6.5		92.4
7.3		27.6
7.5	105.6	
8.5	132.0	

1977). At optimal pH (7.3) the activation energy for nitrite oxidation is much lower than at more acid or alkaline pH values, whereas the activation energy for ammonium oxidation is much less affected by pH. This means that at extremes of pH, the effects of temperature will be more pronounced on nitrite than ammonium-oxidizing organisms, whilst at optimum pH ammonium-oxidizing organisms will be the more temperature sensitive. Wong-Chong & Loehr (1975) further show that the optimal temperatures for the two processes vary with pH.

Heterotrophic nitrifying organisms are a diverse group (chapter 1), which may come into prominence at high temperatures. Since they use organic nitrogen rather than ammonium, their activity can be separated from that of autotrophic nitrifying species by varying the substrates supplied and by using specific inhibitors. In this way heterotrophic nitrifying organisms have been shown to be important in some desert soils, where nitrate can be formed at temperatures in excess of 40 °C and also in the composting of solid organic waste. Such high temperatures do not normally occur in natural aquatic systems. Even in liquid-waste treatment plants, where temperatures may exceed 40 °C, heterotrophic nitrification does not appear to occur, possibly due to the absence of actinomycetes (see discussion and references in Focht & Verstraete, 1977).

Predictive modelling

Attempts have been made to predict rates of turnover of nitrogen and other nutrients from known environmental parameters. One such attempt is designed specifically for tropical forests (Vitousek, 1984). To predict the effects of rainfall and temperature Vitousek defines a climatic index S_v:

$$S_v = (26 - 0.007LAT^2 - 0.0045ELEV)\ln PPT$$

where 26 is the mean annual sea-level temperature at the equator, °C; 0.007 and 0.0045 are coefficients based on the mean temperature gradient at sea-level at latitudes between 0° and 30° N or S and the average of wet and dry adiabatic lapse rates (rates of change of temperature with height with saturated and dry air): latitude (LAT) is in °N or °S and elevation (ELEV) and precipitation (PPT) in metres. Although much of the observed variation could be explained by this index there were significant variations; forest ecosystems will be discussed in chapter 5.

PART II

Case histories from particular environments

4

Nitrogen cycling in dry areas

Introduction

Recent catastrophic droughts have resulted in much publicity for the dry areas of the world as well as the coining of new words such as 'desertification' to describe the spread of deserts. It is now usual to distinguish between arid and semi-arid environments and this practice will be attempted here. Arid environments occupy an estimated 20% of the world's surface and semi-arid a further 15% (Grove, 1985), but these figures are continually changing as a result of natural and man-made processes. Truly arid areas have been defined in various ways, but in general terms they are areas where the energy available (radiation) for evaporation of water exceeds that required to evaporate all the received precipitation. Further, such areas may have one or more years without rainfall and when rains do occur they show wide variations in both space and time. Frequently, arid areas are also hot and saline and some have extreme soil pH values (generally high). Semi-arid areas are usually classed as those with a reasonably predictable rainy season, often short, but usually >250 mm, during which rain-fed crops may be grown.

Not only does rainfall vary in space and time in arid areas, but also vegetation and soil nitrogen pools are very patchy. Thus a wide range of desert ecosystems has been described. Almost all combinations of the various levels of factors listed in Table 4.1 are found and these may be combined with additional problems, including high levels of particular elements such as magnesium. We shall take some examples from these numerous possibilities. The selection is largely a result of the information available, which is generally, like the vegetation of arid areas, very sparse, and does not always reflect global importance. Detailed studies are urgently needed on a range of arid lands if their problems and potentials are to be understood and realised (Wickens, Goodin & Field, 1985).

Table 4.1. *Components of arid areas which may affect nitrogen cycling*

Component	Variation/comment
Precipitation	Frequency, amount, type (rain or snow)
Soil nitrogen	Form, amount, distribution
Temperature	Wide extremes, often in one area (diurnal and/or seasonal patterns)
Salinity	Wide extremes
Alkalinity	Very variable, but often high enough to encourage NH_3 volatilization
Dry deposition	From very small to major movements involving N

Arid areas

Arid areas without vascular plants

Really dry areas, especially those where any rainfall occurs in scattered showers, each of a very small amount, tend not to have vascular plants. This is because growth of vascular plants is related to cell turgor, and maintenance of cell turgor for even a small plant requires a considerable quantity of water. Bacteria do not show turgor as such: they do not have vacuoles. Apices of fungal hyphae have only small vacuoles. These and other properties of the microflora (Griffin, 1981; Skujiņš, 1984) make them more tolerant of low water potentials than higher plants. Not only can they often *survive* long periods of desiccation, but many of them can grow at very low water potentials. Even amongst those which cannot, it takes only a very little water to satisfy a bacterial cell, compared with even one cell of a higher plant. However, bacterial distribution may be constrained by the number of water-filled pores through which individuals can move. Filaments (actinomycetes, fungi) can cross air-filled gaps and so are particularly significant in dry areas. These considerations suggest that unusual nitrogen cycles may occur in dry regions. Figure 4.1 illustrates one such system, that in the cryptogamic crust found in many desert areas, including the winter semi-deserts of the USA (West and Skujiņš, 1977), but not where there is significant sand movement (Skujiņš, 1981). An essential feature of most such crusts is a nitrogen-fixing cyanobacterium which may be free-living or part of a lichen symbiosis. In either case the organism is able to withstand long periods (years) of desiccation and yet rapidly (minutes) resume metabolic activity on rehydration. As light energy is usually in excess of that required to saturate photosynthesis, photoautotrophy is an obvious advantage. Desiccation-tolerant mosses may be included in the crust. Overall, these crusts tend to be rather stable and may have a significant effect in preventing soil erosion: their C:N ratio is rather low, largely due to the cyanobacterial compo-

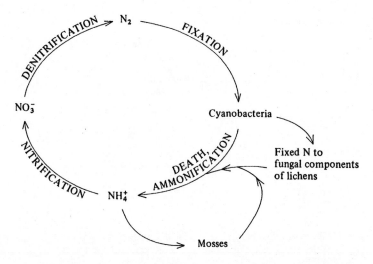

Figure 4.1. Nitrogen cycling in a cryptogamic crust of a desert. After West & Skujiņš (1977).

nent. This, together with the prevailing temperatures, results in high rates of microbial respiration beneath the crust, generating anaerobic microsites where denitrification is rapid. Thus, although nitrogen fixation rates may be locally high, nitrogen accumulation below crusts is usually limited. This may be why most deserts have little soil nitrogen.

High nitrate levels are occasionally found (Charley & McGarity, 1964): these may result from repeated series of events as follows:

rain → mineralization of litter → nitrification.

Each rain shower rehydrates sufficient numbers of microorganisms to allow some mineralization and nitrification, so that after each shower the soil becomes enriched in nitrate. A gradual increase in levels of nitrate and associated cations occurs. For this to continue to a high level, the rainfall at any one event must be low enough to (a) prevent leaching of nitrate, although there may be movement down to the leaching boundary and (b) not to allow the growth of vascular plants, because these would deplete the soil of nitrate. It is obviously further necessary that conditions should not favour denitrification.

Introduction of a spermatophyte component

Desert vascular plants are mainly spermatophytes and fall into two broad categories, ephemerals and perennials. For ephemerals to persist, sufficient rain must fall and be retained in the soil to allow a complete life cycle from seed to seed. Such a rainfall event need not occur

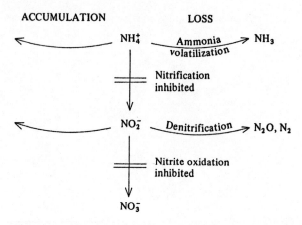

Figure 4.2. Possible effects of some desert shrubs on nitrification, leading to either nitrogen loss or accumulation of intermediates. For details see Skujiņš (1981).

every year, but equally could occur more than once in any particular year. Desert perennials are usually either geophytes, succulents or woody shrubs. Geophytes are generally subject to the same constraints as ephemerals, requiring a rainfall event, not necessarily every year, but of sufficient magnitude to allow a cycle from and to the vegetative subterranean perennating structure. Succulents show extreme water conservation plus physiological adaptations such as crassulacean acid metabolism. Desert shrubs are versatile, often having photosynthetic stems and the ability to produce and shed leaves very rapidly, usually coupled with the water conserving C_4 carboxylation system.

All of these spermatophytes require nitrogen, but the details of nitrogen cycling vary, especially in the proportion of total ecosystem nitrogen in the biomass. Where ephemerals predominate, this also varies greatly in time, from most of the nitrogen being in the biomass during plant growth, to very little in the dry season. In the case of perennials, all retain a high proportion of total ecosystem nitrogen throughout the year, and carry out considerable internal recycling. External cycling involves but a small part of the total nitrogen, which, in the case of succulents with ephemeral roots, is largely underground. Further, some desert perennials have marked allelopathic effects on the soil flora. Three such are described by Skujiņš (1981). First, *Atriplex* and *Artemisia* species may repress growth of cyanobacteria, thus inhibiting what is probably the major nitrogen-fixing component in their locality. Second, these and other higher plant species may inhibit nitrification. This can lead to ammonium accumulation if the necessary cation-binding sites are available, or ammonia loss by volatilization, especially at higher pH values and temperatures. Third, *Atriplex, Artemisia* and *Ceratoides* species may inhibit the oxidation of nitrite to nitrate during nitrification. This can lead to nitrite accumulation, but more often, especially when sufficient organic matter is available, nitrite is diverted to denitrification processes. These points are summarized in Figure 4.2. It appears that some desert shrubs may have a very adverse effect on soil nitrogen by the combination of inhibition of nitrogen fixation and enhancement of nitrogen loss. The benefit to the plant of these effects is difficult to see, except in competition terms if they have a particularly efficient system for scavenging soil nitrogen.

The role of legumes in desert nitrogen cycling

Many leguminous shrubs and small trees grow in desert areas and it is often assumed that this is when (as often) soil nitrogen levels are low, making the ability to fix nitrogen a competitive advantage. Unfortunately, evidence, rather than supposition, is largely lacking. A number of drought-tolerant legumes (e.g. *Cercidum*) cannot nodulate. Others (such as most *Acacia* and *Prosopis* spp.) which can nodulate have not been shown to do so in truly arid areas (Sprent, 1985, 1986a, 1987). It is not sufficient to see even a potentially nodulated legume growing quite well to assume that it is fixing nitrogen. Woody legumes from arid areas generally have one or both of the following attributes: (a) very deep roots, (b) adaptations such as phyllodes which minimize water loss. These alone may lead to their success in dry areas. Indeed, to make a good root system to acquire nutrients in general, including any available water, is a much better use of plant resources (particularly nitrogen) than to make root nodules (Sprent, 1985). There is only one case (discussed below) where there is evidence to suggest that shrubby legumes may fix nitrogen in desert areas.

Deserts with deep water tables

Deserts may have deep water tables, for example, parts of the Sahara (although the water table is dropping due to removal from wells) and much of the Sonoran desert in Arizona/Mexico. The latter is more of a semi-desert since it occasionally has very heavy rainstorms and flash floods which cause water erosion. Phreatophytes are plants with very deep roots which can grow down to such water tables. Some species of the mimosoid legume genus *Prosopis* fall into this category, which is why they

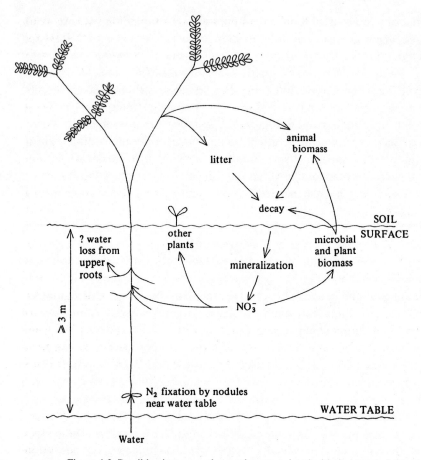

Figure 4.3. Possible nitrogen cycle reactions associated with deep-rooted desert mimosoid legumes such as *Prosopis* spp.

grow well in environments where the upper soil layers and the atmospheric conditions may be typically desert-like. In the vicinity of such *Prosopis* plants, there may be an interesting variant on the nitrogen cycle as follows: roots of *Prosopis* near the water table nodulate and fix nitrogen to support its growth. *Prosopis* biomass either directly (via litter) or indirectly (via herbivores) returns nitrogen to the surface or upper soil layers. Here mineralization takes place and nitrate may accumulate. Because the nodulated roots do not have access to this nitrate (there is insufficient water to wash it down to them) nitrogen fixation is not nitrate inhibited. Thus, depending on the extent of denitrification in relation to nitrogen fixation, soil nitrogen levels may steadily build up. Other, non-nitrogen-fixing plants may be supported by this nitrogen, subject to the overall constraint of limiting water supply. Even this may be partly offset by the phreatophyte, since recent evidence supports earlier, contentious, hypotheses that water may pass from deep water tables, through lower roots and out of upper roots (e.g. Bavel & Baker, 1985). This water may support microorganisms, including species taking part in the nitrogen cycle. Figure 4.3 summarizes these possibilities. Deep-rooted plants, many of which do not fix nitrogen, may also bring within the root zone of other plants various nutrients, including combined nitrogen dissolved in groundwater. Much of this groundwater has, like oil, been present for millions of years and is not being replaced. It should be (but usually is not) carefully husbanded. One attempt to minimize water loss and protect deep water supplies involves the killing of deep rooted shrubs with herbicides and encouraging shallow-rooted grass species. This method is only feasible where there is sufficient rainfall to support the grass, such as in mountainous chaparral areas of Arizona, USA (Davis & Debano, 1986). The procedure was successful in that more water reached lower areas, but this water had up to 6 mol m^{-3} nitrate compared with down to 3 mmol m^{-3} in untreated areas. This nitrate, which was formed following mineralization of the large quantities of dead below-ground biomass, would be useful for irrigating crops but is above the recommended levels for drinking water.

Erosion losses of nitrogen from arid areas

Erosion may occur in one of two ways, by water or by wind (also known as aeolian erosion), and it may result in large-scale redistribution of soil nitrogen. In deserts such as the Sonoran, where flash floods may follow sudden rainstorms, between 0.2 and 25 kg N ha^{-1} may be lost in a year (Skujiņš, 1981). In this area aeolian losses are much lower – the famous tall saguaro cacti (*Carnegia gigantea*) would not remain upright in excessive wind! Elsewhere, for example in North African deserts, aeolian losses are the greater.

Can the nitrogen cycle be manipulated to halt and reverse desertification?

This is the subject of much current effort. Large-scale irrigation has produced almost as many problems as it has solved (Wickens *et al.*, 1985). In most arid areas the peoples are poor, precluding extensive use of fertilizer nitrogen. It may also be pointless to fertilize unless suitable slow-release formulations are developed and used, because of gaseous losses of nitrogen. For example, urea nitrogen may be lost as ammonia and nitrate nitrogen by denitrification to nitrogen gas. For further discussion of factors affecting gaseous loss of nitrogen from these and other areas, see Freney & Simpson (1983). Introduction and protection of woody species (which are at a premium as fuel for cooking) offer considerable scope for amelioration of deserts, because nitrogen and water are retained in the biomass and also because they have beneficial effects on the microenvironment (this of course would merely return these areas to the condition they were in previously).

The role of animals

As mentioned briefly in chapter 2, termites and other arthropods are the most important invertebrate metazoans in dry areas, whereas annelids have an important role in wetter areas. As elsewhere, their role is twofold – acting as redistributors of nitrogen and as accelerators of nitrogen cycling. These roles may be particularly important in dry areas. by taking plant nitrogen from dry surface to moist lower layers, the chances of mineralization are increased. Further, desert arthropods are well designed to minimize water loss and their gut flora and fauna are provided with a warm, moist environment in which nitrogen may be mineralized even when soil microorganisms have lost metabolic activity due to insufficient moisture.

There may be interesting variations within animal groups. For example, two types of termite occur in arid areas, those which form mounds and those which do not. In the mound-forming species (Figure 4.4A) workers bring plant material into the mound for consumption by other castes. As a result of digestive processes, the recycling of nitrogen by consumption of cast skeletons, dead and sometimes living termites, and the use of faeces for mound construction, the mound becomes gradually enriched in nitrogen. Since no nitrogen is removed (although the possibility of some denitrification cannot be excluded) and carbon dioxide is lost by respiration, the C:N ratio falls. Mound-forming colonies of termites may live in excess of 100 years. At death the mounds disintegrate and a locally rich nitrogen source is available for plant and microbial growth. In parts of Asia termite mounds are spread as fertilizer.

Termites which do not form mounds take plant material into the soil where the nitrogen contained in it is eventually mineralized. Use can also be made of this process. In the Sudan, Dinka tribesmen (in the past, when trees were still available) felled trees and shrubs and laid them in piles.

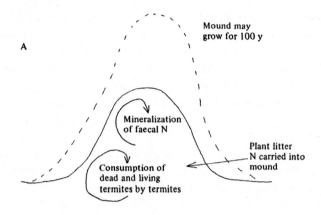

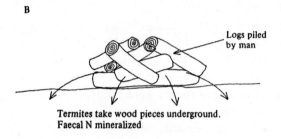

Figure 4.4. Effects of termites on mineralization of nitrogen. A, mound-forming termites; B, non-mound-forming termites.

These piles were attacked by termites and the nitrogen in them transferred to the soil (Figure 4.4B), much more rapidly than would have been the case in the absence of termites (Lee & Wood, 1971). Other pressures on wood (most notably for fuel) may now have superseded this use.

Semi-arid areas

Semi-arid areas fall into many types, depending on the season during which the rain falls. At one extreme is the Mediterranean pattern, found not only in the Mediterranean region, but also in many other places such as parts of southern Africa and Western Australia. Cool, wet winters are followed by hot, dry summers. At the other extreme are those regions of Australia, Nigeria, Argentina and elsewhere which have their rainy period during the summer, and cool, dry winters. Within these groups are various sub-groups in which one of the main variables is temperature: in general these are more extreme away from coasts. There are also intermediates between these rainfall extremes, for example the Wagga Wagga area of Australia where the rainfall is distributed more-or-less evenly throughout the year. Some examples of nitrogen cycling from these types will be given.

Probably the most widely studied of all the arid and semi-arid areas of the world are those with a Mediterranean climate. There is a long (several thousand years) history of agriculture in the vicinity of the Mediterranean Sea. Because of the economic importance of this agriculture, studies have mainly been directed at improvement of levels of nitrogen in the system so as to optimize crop and livestock production (see, for example, the various papers in Monteith & Webb, 1981), rather than to understand nitrogen cycling. Recently, interest in better use of semi-arid rangelands for agriculture has led to investigation of natural nitrogen cycling in these regions, and more definitive data should soon be available. Because the rain occurs in the cool season, low temperature is sometimes more limiting than water deficit to biological processes during this, the main growing season. All reactions of the nitrogen cycle are likely to proceed broadly as in moist temperate areas during the wet season. Volatilization losses at this time are likely to be small, but to increase with the onset of the hot dry season. Because of limited rainfall, leaching losses are small.

A very detailed study of trophic interactions and nitrogen cycling in a semi-arid short-grass prairie in Colorado has highlighted the relations between microorganisms, soil invertebrates and plants (Ingham *et al.*, 1986a). In this particular situation, rainfall is scattered throughout the growing season, but soil moisture is lowest in summer due to high evaporation. Once the components of the flora and fauna were described and enumerated (no mean task!), perturbations were introduced. These included, for example, the use of nematocides to remove the nematode component. Results show that this particular system is relatively stable to abiotic fluctuations. One reason for this is thought to be the ability of the predators in the ecosystem to use more than one prey group. Within the systems which have been studied, this type of stability appears to be unusual.

Studies of nitrogen cycling in natural and cropped semi-arid areas of the Australian Northern Territory, which has summer rainfall, have been

Table 4.2. *Some symbiotic nitrogen-fixing plants found following loss of nitrogen by wild fire. The first three categories usually grow from seed, the last regenerates from boles*

Plant type	Example of location
Pasture legumes	Australian Northern Territory
Woody legumes (e,g, *Acacia* spp.)	Western Australia
Actinorhizas (e.g. *Ceanothus* spp.)	West Coast of USA
Cycads (e.g. *Macrozamia reidlii*)	Western Australia

proceeding for a number of years. Much of the work has been summarized by Wetselaar (1980). Typical vegetation is a savannah woodland with *Eucalyptus* spp. as dominant trees and *Sorghum plumosum, Themeda australis* and *Crysopogon fallax* (all C_4 grasses) as dominant ground flora. The annual turnover of these is estimated respectively to be 770 and 1500 kg ha^{-1} dry weight and 8 and 16 kg N ha^{-1}. The total nitrogen turnover in this system is thus relatively small and animals appear to play a major role in it. The most immediately obvious animals are the marsupials (mainly kangaroos and wallabies), but their density is estimated to be about 1 kg ha^{-1} liveweight, compared with 700 times this for termites. At known rates of dry matter consumption, 700 kg ha^{-1} termites could consume *all* the available plant material. Termites of this area are mainly mound-formers, which, as discussed earlier in this chapter, form local closed units of nitrogen cycling. Assuming that a proportion of these mound colonies dies each year, there is likely to be a regular but patchy input of readily available nitrogen.

Wild fire is a frequent occurrence in this area and may lead to losses of >90% of the nitrogen in pasture and leaf litter. When this occurs, the nitrogen cycle becomes biased in favour of nitrogen-fixing species – in this case principally pasture legumes (alternatives which may be found in other systems are listed in Table 4.2). Pasture growth in the wet season following a fire may be reduced by 50%, with associated effects on termites, until the nitrogen-fixing species (which may include some non-symbiotic forms) build up soil nitrogen to its previous levels. During the wet season, remaining plant litter decomposes very rapidly, even though it has a low %N, because high temperatures and high humidities allow rapid microbial growth. Some of the plant nitrogen passes from above-ground to below-ground pools, where it is protected from loss by burning, as a result of translocation within plants and leaching from trees (throughfall and stem flow) during the wet season. As Wetselaar (1980) points out in summarizing this work, little was known about gaseous losses of nitrogen at the time the measurements were made. Such losses may contribute to the overall nitrogen cycle in this, as in other systems. In any case, the components of the nitrogen cycle can be seen here to vary in a fairly regular seasonal way, upon which is superimposed a less-well-defined temporal pattern due to fire (which varies not only in frequency, but also in intensity) and a spatial pattern associated with location of components such as trees and termite mounds, to say nothing of topography. Taken together, these factors make many semi-arid environments rather fragile and in need of careful management. The example given here is probably typical in general terms of many semi-arid savannah grasslands. Medina (1982) compares it with similar areas in Africa (Nigeria and the Ivory Coast) and Venezuela.

The main agricultural use of savannahs such as that described above is for cattle grazing. Cattle themselves (providing stocking rates are not too high) have little overall effect on nitrogen cycling compared with termites, because, as with marsupials, their relative liveweight per unit area of land is low. However, if land is cleared, rather than grazed in its natural state, then the nitrogen cycle is greatly disturbed.

Effects of clearing semi-arid areas

These effects have some highly characteristic features, so they will be considered here rather than in chapter 7. Again the main example will be from the Australian Northern Territory (Wetselaar 1980). The clearing process consists of felling and burning of trees, followed by ploughing. The soil may then be left fallow (unplanted) for one or more years. In the wet season very rapid mineralization of soil nitrogen occurs, with most being oxidized to nitrate. This moves up the soil profile at the end of the wet season as soil dries, and down the profile at the onset of the next wet season. Clearly, the rooting pattern of subsequent crops will be important in determining how the nitrogen cycle proceeds from here. In this particular system, soil organic nitrogen is relatively high (c. 1600 kg ha^{-1}) and only a part of this is mineralized each year. However, rates of mineralization may drop with time as the nature of the organic material changes. This is clearly shown in the history of farming in another part of Australia, the brigalow country of Queensland. The original flora of this area had as its dominant tree *Acacia harpophylla* (commonly known as brigalow). This plant fixes nitrogen, and the soil under it at the time of clearing was very rich in nitrogen. After clearing, mineralized nitrogen supported good growth of pasture grasses for about 10 years and then there was a rapid decline. The decline could be halted by use of fertilizer nitrogen or by the inclusion of a legume break crop. Thus in these semi-arid areas, as in very wet peat systems, it is the available nitrogen, rather than the total nitrogen, which is important to the rate of nitrogen cycling (Graham, Webb & Waring, 1981).

Dry, infertile forests

These occur in various areas. One where the nitrogen cycle has been studied in some detail is in south-eastern Wyoming, USA (Fahey, Yavitt, Pearson & Knight, 1985). The dominant tree here is lodgepole pine, *Pinus contorta* ssp. *latifolia*. Winters are long and cold and about 400 of the 600 mm annual precipitation falls as snow during this period. Summers are short and cool. The system studied is still aggrading more than a century after a forest fire. As with other such oligotrophic sites (see chapter 5 for an example from a tropical rain forest), nitrogen retention is high. Very little leaching occurs although some nitrogen may be removed laterally by soil arthropods and fungal hyphae (particularly mycorrhizal species). Figure 4.5 summarizes the major nitrogen components and flows. Most of the nitrogen is in the soil organic fraction. At any one time inorganic nitrogen accounts for 2 to 6 kg ha^{-1}, mainly in the form of ammonium. However, conditions for oxidation are good and nitrifying bacteria present, so that nitrification cannot be ruled out if it is coupled to rapid uptake by roots and/or heterotrophs. Further, nitrate and ammonium (the amount of nitrate usually exceeds that of ammonium) enters the soil by precipitation and throughfall. The latter also includes organic material. However the total of these was below 3 kg ha^{-1} and, of course, the throughfall component is not a net input. Heterotrophic nitrogen fixation occurs in litter and there are patches of *Lupinus argenteus* showing symbiotic fixation. The total fixation component is likely to be well below 1 kg $ha^{-1} y^{-1}$. Thus inputs are low and outputs even lower, resulting in a slow, but steady accumulation of nitrogen. No information is available for gaseous nitrogen loss, but the soil and climatic conditions obtaining make this unlikely to be great. Apart from this, a complete nitrogen cycle seems to operate, but because of low capital and high demands, mineralized nitrogen is utilized very rapidly. Leaching by snow melt and rainfall is insufficient to allow major losses to groundwater.

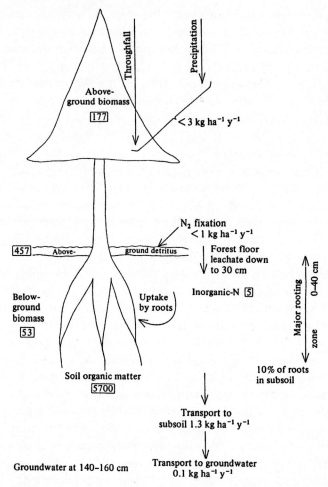

Figure 4.5. Major pools of nitrogen in a lodgepole pine forest in Wyoming, USA. From Fahey *et al*. (1985). Figures in rectangles are in kg ha^{-1}. Other figures are annual nitrogen movement.

Coastal sand dunes

These may also be limited from time to time by water supply. The example to be taken is from Lincolnshire, England (3°21′N, 0°15′E) (Skiba & Wainwright, 1984). The major features of the uncolonized sand and that under *Ammophila arenaria* and *Hippophäe rhamnoides* are given in Table 4.3. The levels of ammonium and nitrate are quite high – they correspond to concentrations of approximately 3 and 2 mol m^{-3} ammonium and nitrate respectively for the climax soil at water holding

Table 4.3. Nitrogen content and pH of coastal sand dunes in Lincolnshire, England (from Skiba & Wainwright, 1984). For comparison, some data from a Pinus contorta forest are included (Fahey et al. 1985)

| Area | pH | g N g^{-1} soil as | | C:N | % of total N as NH$_4^+$ + NO$_3^-$ | Soil organic matter (%) |
		NH$_4^+$	NO$_3^-$			
Uncolonized sand	8.55	15.3	6.6	7.6	64.4	0.05
Rhizosphere of Ammophila arenaria	8.37	16.3	7.6	8.9	37.9	1.2
Rhizosphere of Hippophäe rhamnoides	8.45	16.3	6.3	5.9	36.3	0.07
Soil from mature (climax) dunes	7.83	16.7	10.4	53.5	27.1	1.1
P. contorta forest soil	acid			40.5	0.1	7.9

capacity. However, both forms of nitrogen, but particularly nitrate, are readily leached, so that nitrogen can be a factor limiting plant growth. The balance between ammonium and nitrate was thought to be strongly conditioned by pH. Laboratory studies with added ammonium and nitrite led to results which were consistent with the hypothesis that very alkaline pH values, coupled with high levels of ammonium, inhibit nitrite reduction, but that at pH 7.83 this can proceed more rapidly. Further, nitrate was reduced rapidly to ammonium in dune soil. The consequence of this should be accumulation of nitrite at some sites, but although this occurred in laboratory studies, it was not found in the field. This emphasizes the need for coordinated field and laboratory studies.

In the rhizospheres of *A. arenaria* and *H. rhamnoides* and in the uncolonized sand, the C:N ratio is very low, such as would permit rapid cycling of nitrogen through inorganic fractions and little immobilization. In the climax system it is much higher, largely due to a higher organic carbon content, and only about 25% of the nitrogen is in inorganic form (ammonium + nitrate). Overall, dune systems have a very variable water status, in which leaching may be important at some times and absent at others. Together with high pH this may lead to rapid cycling of nitrogen.

The sand dune system described in this section and the infertile forest system described earlier represent two very different systems operating on a limited water supply. Table 4.3 shows some of these differences. Both are aggrading systems. The sand dune system starts from a very low nitrogen level, the forest much higher (after the fire which was taken as the base line). However, available (inorganic) nitrogen as a percentage of total nitrogen is very low in the *Pinus* system. Although direct comparisons are not possible, the total available nitrogen (ha^{-1}) may not be very different. At both mature sites the C:N ratio is such that nitrogen is likely to be immobilized. In both, ammonium is largely bound. The dune soils, however, because of their free drainage and frequent rain are more likely than the forest to have leaching losses.

5

Terrestrial areas not subject to regular drought

Introduction

This chapter considers all those areas where water supply is not normally a problem except in a waterlogging sense. Rainfall varies from low (high latitudes, low evaporation) to very high. The examples taken will follow a generally latitudinal sequence, beginning at the poles.

Tussock tundra

Tundra vegetation is characteristic of certain arctic areas of latitudes above 60 °N. It varies considerably in rainfall, evaporation and in vegetation, particularly the balance between angiosperms and cryptogams. The example taken is from Alaska, a type of vegetation known as tussock tundra. Salient features are indicated in Figure 5.1. By the addition of soluble carbon (in the form of corn starch) and/or soluble nitrogen (in the form of urea, labelled with ^{15}N in some cases) at the beginning of the growing season, competition for nitrogen within the system was assessed. In peaty systems of this type, the living vegetation consists mainly of vascular plants and mosses and contains only about 1–2% each of the total organic matter, total nitrogen and total phosphorus, with the balance in the non-living parts, i.e. litter + brown moss + soil. This contrasts with most tropical systems in which the bulk of the organic matter is in the living biomass. On addition of ^{15}N labelled urea to the tundra, two distinct patterns of uptake were seen. In mosses (mainly *Sphagnum* spp.) there was a rapid uptake immediately after application and little further uptake, whereas the vascular plants took up ^{15}N throughout the growing season. This reflects the different soil sources available to these two categories of plant. Mosses, having no true roots, can only utilize nitrogen at or near the surface, whereas vascular plants can exploit the whole volume of soil through which their roots ramify. However, 70–80% of the added urea nitrogen was incorporated into the

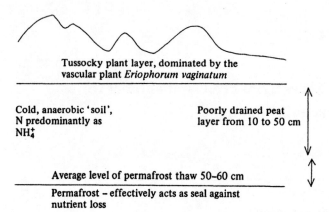

Tussocky plant layer, dominated by the
vascular plant *Eriophorum vaginatum*

Cold, anaerobic 'soil',   Poorly drained peat
N predominantly as   layer from 10 to 50 cm
NH_4^+

Average level of permafrost thaw 50–60 cm

Permafrost – effectively acts as seal against
nutrient loss

Figure 5.1. Major features of an arctic tussock tundra system, studied by Marion, Miller, Kummerow & Oechel (1982). The growing season is from June to August, when the mean air temperature is 8–16 °C. Annual rainfall is about 150 mm.

insoluble soil fraction and the size of this fraction was increased when corn starch was added. Some interesting inferences may be drawn from these data. It is clear that the microbes are normally limited by soluble carbon, and when any is available, these same microbes sequester soluble nitrogen. This leads to a delicate balance between soluble forms of these two elements. If the vascular plants are nitrogen, rather than carbon limited, then they should be able to compete successfully for soil nitrogen with the microflora which is carbon limited. It was thus concluded from this study that vascular plant growth was limited by the amount of nitrogen made available by soil mineralization processes. The major input of nitrogen into the soil is likely to be through the surface via precipitation and snow melt, by leaching of live biomass and litter. Spatially, these inputs are all available first to the moss flora, which is thus unlikely to be nitrogen limited. Further, ammonium may be physically bound to *Sphagnum*.

In some wet tundra areas, cyanobacteria (particularly species of *Nostoc*) fix nitrogen; these cyanobacteria may be associated with mosses. At slightly drier sites, lichens take on a greater role, including, in appropriate species, nitrogen fixation (Cleve & Alexander, 1981). At even drier sites, nitrogen fixation is depressed. Losses of nitrogen by denitrification are probably very low because of both the low temperature and the lack of nitrate. A summary of the nitrogen cycle in wet tussock tundra is given in Figure 5.2. Note that it is very similar to that obtaining in many forest systems, but with some important differences (Table 5.1). The rate of

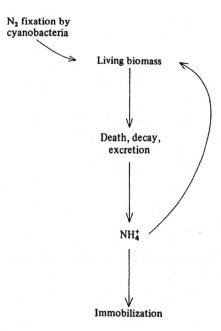

Figure 5.2. Nitrogen cycling in a wet tundra ecosystem.

Table 5.1. *Comparison of nitrogen cycles in arctic tussock tundra and some temperate forests*

Component	Tundra	Forest
Nitrification	Largely precluded by environment	May be present or precluded by nitrification inhibitors or unsuccessful microbial competition for ammonium or soluble organic-N
Denitrification	Largely precluded in absence of nitrification	
Living N in biomass	Mainly cryptogamic	Mainly in trees
N in microbial biomass	Low	Higher
N fixation	Cyanobacteria (lichens)	Lichens, free-living heterotrophs

nitrogen cycling is usually slow because of low temperatures. It may also be limited by other nutrients because of poor weathering of rock, especially where the rock is under a peat layer in the permafrost. Thus phosphorus may limit productivity in parts of Alaska. Although soluble plus extractable pools are small, turnover of nitrogen and phosphorus have been estimated as 11 and 200 times per growing season – values usually associated with the tropics (Bunnell, Maclean & Brown, 1975). The relative roles of biological and abiological processes in these rates are not yet clear.

As elsewhere, animals are important in Arctic tundra systems (see volume edited by Bliss, Heal & Moore, 1981). They range from protozoa to vertebrates including lemmings (Rodentia). When lemming populations reach their famous periodic heights, effects on standing biomass and nitrogen return to the surface (both as felled biomass and faeces) are great. Because plant material in tundra generally has a low concentration of nitrogen, many vertebrate herbivores have nitrogen-conserving systems such as recycling of urea nitrogen (via microflora) in ruminants.

Forests

Where rainfall is not limiting and temperatures not too low, climax vegetation tends towards forest. Many studies on nitrogen cycling in forests have been conducted and many generalizations made. Predominant amongst these are (a) that there is more litter in high than low-latitude forests and (b) that high nitrogen concentrations in soil lead to high nitrogen concentrations in leaves, which in turn lead to high rates of nitrification and results in a mull type of humus: there is a parallel series for soils with low concentrations of nitrogen, ending with a mor humus. These sequences of events affect the proportioning of nitrogen between the plant and microbial components (Figure 5.3). Further, broad leaves are thought by some to break down more readily than needles. These arguments have been challenged by Vogt et al. (1986) who suggest that evergreenness (usually associated with litter of about 20 Mg ha^{-1}) or deciduousness (litter about 10 Mg ha^{-1}) of foliage is more important than latitude or whether foliage is in the form of needles or broad leaves. However, forests at latitudes above 40° tend to have more forest floor litter than those below 40°. Additional factors coming into play include leaf quality, especially lignin content, because lignin tends to retard decay (Figure 2.4), and nitrogen content. Herbivory tends to be associated with high nitrogen concentrations in leaves and high nitrogen availability. If combined nitrogen is plentiful, then loss of some of it by herbivory is not

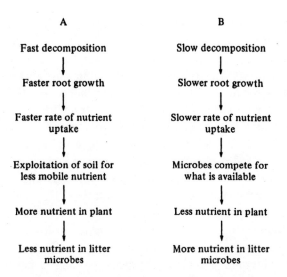

Figure 5.3. Two hypothetical sequences of nutrient turnover in forest soils. (A) occurs when there is a plentiful supply of nutrients (including nitrogen), (B) when nutrient supplies are low. These sequences are discussed and challenged by Vogt *et al*. (1986).

a great disadvantage to the plant; if it is scarce, diversion of energy to discourage herbivores may be advantageous.

In forest (and other) studies, root biomass is frequently ignored or estimated only approximately. As Vogt *et al*. (1986) point out, this may lead to gross errors. For example it was calculated that, except for cold-temperate broad-leaved forests, root turnover added 18–58% more nitrogen to soil than litterfall. Further, inclusion of roots leads to a considerable decrease in the estimated mean residence time of organic matter. Bearing these variations and problems in mind, some examples of forest nitrogen cycling will be given.

Evergreen forest, dominated by Eucalyptus

In much of Australia, the broad-leaved evergreen genus *Eucalyptus* dominates. Adams & Attiwill (1986a,b) studied sites in Victoria having various levels of rainfall. As in all evergreen systems, litterfall is continuous, but shows peaks at certain times of the year (Lamb, 1985). Litter of *Eucalyptus* species consists not only of leaves but also considerable quantities of wood (twigs, etc.) and bark (bark peeling being a characteristic feature of many eucalypts). Most eucalypts grow on highly leached sandy soils with few nutrients. They have considerable capacity for conserving and internally recycling nutrients, including nitrogen: indeed, most of the nitrogen in such systems is in the living tree biomass (Vogt *et al.*, 1986). Leaves when shed contain only about 0.9%N and yet the mean residence times for leaf nitrogen in litter are on the low side (0.7–1.5 years, compared with 2.0–6.8 years for all litter). The rates of reaction of the individual processes of the nitrogen cycle vary greatly with site and season. Some of these variations are given in Table 5.2 for sites with four different *Eucalyptus* species and varying temperature and rainfall patterns. Much of the variation in nitrogen mineralization and turnover can be attributed to rainfall and temperature. Thus in both the *E. regnans* and *E. microcarpa* sites, which are respectively cooler and wetter and drier and warmer, about 50% of the mineralizable nitrogen is nitrified, whereas nitrification on the other two sites is insignificant. In the *E. regnans* site, mineralization always exceeds immobilization, so that ammonium is always available. The overriding factor in the *E. microcarpa* site may be the low C:N ratio of the soil, which allows some mineralization to proceed in parallel with immobilization. The %N in leaf litter is higher than in some plots largely because of an understorey of nitrogen-fixing plants such as *Acacia* species. For *E. sideroxylon* the mean residence time for nitrogen is very low although nitrification is insignificant, possibly due to a high C:N ratio. In earlier work on these soils the authors found that nitrification did not proceed above a C:N ratio of 18–20. For this reason, ammonium is likely to be the main form of nitrogen both produced in soils and taken up by plants. Adams & Attiwill (1986a) used the climatic index of Vitousek (see pp. 65–6) to investigate the overall pattern of nitrogen turnover. Although this index was derived for tropical forests, it fitted the *Eucalyptus* data even better! Applying it led to the conclusion that litter from forests in moist areas (tropical or temperate) tends to have a short residence time, whereas that from drier and/or colder (high elevation) areas has a longer residence time.

Eucalyptus spp., even when growing on moist sites, are prone to fire and this can have an overriding effect on nitrogen cycling. This may lead to nitrogen loss to the atmosphere (p. 9) and also encourage growth of nitrogen-fixing plants (p. 79). In the study described above, fire occurred on the *E. regnans* and the *E. obliqua* sites. In both cases, immediately following fire, the potentially mineralizable nitrogen doubled and the available nitrogen rose to values equivalent to the potentially mineralizable nitrogen before the fire. These peaks were short-lived and nitrogen immobilization resulted in values dropping to pre-fire levels within 6

Table 5.2 *Some characteristics of* Eucalyptus spp. *forests studied by Adams & Attiwill (1986a)*

Characteristic	Species of *Eucalyptus*			
	regnans	*obliqua*	*sideroxylon*	*microcarpa*
Soil C:N	14.5	23.2	21.1	16.3
Soil pH	4.7	5.1	4.6	4.9
Mineralization(M)/immobilization(I)	M > I		I often ≥ M	
Forest type	Wet sclerophyll	Wet/dry sclerophyll	Dry sclerophyll	
Annual rainfall (mm)	1250	930	570	570
Soil temperature range (°C)	7–17		12–23	
%N in leaf litter	0.80	0.73	0.79	0.89
Nitrification as a % of mineralization	50	Insignificant		50

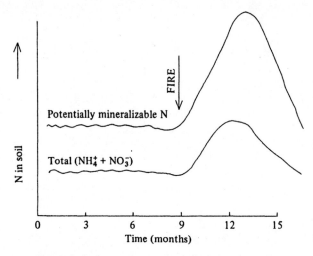

Figure 5.4. Effect of fire on nitrogen mineralization in some *Eucalyptus* forests. After Adams & Attiwill (1986b). Potentially mineralizable nitrogen is defined as the inorganic nitrogen which is extractable in hot potassium chloride.

months (see Figure 5.4). No evidence of nitrogen loss by leaching after nitrification was obtained.

Tropical rain forest

In the wetter areas of tropical rain forest, there is considerable loss of water and solutes, which enter the major river systems such as the Amazon basin. Soils around the Amazon vary greatly and nutrient cycling varies accordingly. Two examples will be taken, one specific and one more broad-ranging.

The first site is in southern Venezuela nearly 2 °N of the equator, with an average annual temperature of 26.2 °C and an average annual rainfall of 3500 mm, with no month having less than 100 mm (Herrera & Jordan, 1981). The soil has low fertility, is very sandy and heavily leached. It supports a type of forest known as the Amazonian caatinga and has over 300 tree species. Only three of these are co-dominant, *Eperua leucantha* (a legume of questionable nodulation status), *Manilkara* sp. and *Micrandra spruceana*. Compared with other tropical forests, the nitrogen contained in litter is high, but by far the largest nitrogen component is in the root fraction (Figure 5.5). Total nitrogen in the ecosystem is only just over 2 Mg ha^{-1} and nitrogen and other nutrients are carefully conserved. Thus, for example, nitrogen is translocated from leaves to the parent plant prior to abscission, resulting in a drop in %N from 1.02 to 0.74, even

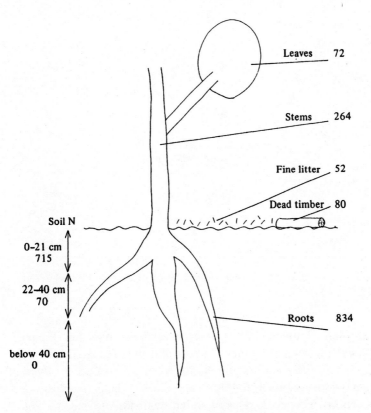

Figure 5.5. Nitrogen pools in an Amazonian tropical rain forest in Venezuela. All figures are in kg N ha^{-1}.

lower than in the eucalyptus forest described above. Since leaves have about 45%C, typical of non-salt-accumulating species, a consequence of this is a high C:N ratio. Much of the carbon is used for strengthening material and/or producing substances to discourage herbivores, rather than in photosynthetic machinery (which also requires nitrogen). However, in humid tropical conditions fast growth of microorganisms can still give rapid mineralization as in the eucalypt system described previously (see also Kawama, 1981). However, this does not necessarily proceed to nitrate for reasons which are not yet clear. In the Amazon system, little nitrate, but more ammonium is found in soil and leachate entering the river. Further, there is a net loss of organic nitrogen, suggesting that mineralization is not complete. This organic material gives the water a characteristic dark colour, hence the name 'Rio Negro'.

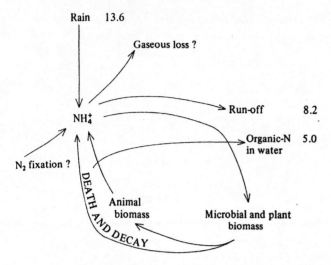

Figure 5.6. Nitrogen cycle for the tropical rain forest illustrated in Figure 5.5. The figures are in kg N ha^{-1} y^{-1}.

From these data a simplified nitrogen cycle can be constructed (Figure 5.6). It can be seen that the input via rain and the estimated loss are approximately balanced. Is nitrogen fixation needed? It can certainly be measured, in the soil humus + root mat, sand bark + lichens and leaf + epiphyte fractions. However, rates of activity vary widely in space and time and are from indirect (acetylene reduction) assays only, so it is impossible to put even an approximate figure on the annual input from this source. Herrera & Jordan (1981), realizing the possible errors involved, suggest a value between 35 and 200 kg ha^{-1} y^{-1}. Since no direct measurements of gaseous losses have been made, a corollary of these estimates for nitrogen fixation is that a similar amount of nitrogen must be lost as gas, as this is a climax, rather than an aggrading system. Since soil pH is low, ammonia loss is unlikely. Waterlogged conditions might favour denitrification although this is likely to stop at N$_2$O (see chapter 3) at the pH levels obtaining. However, a prerequisite for denitrification is nitrate and this appears to be scarce in the caatinga system. We thus have several possibilities (not mutually exclusive) including (a) rapid denitrification of any nitrate formed, so that free nitrate is not detected; (b) grossly overestimated rates of nitrogen fixation; (c) errors in the estimates of input and output by rain and run-off. Of these options (b) is likely to account for the largest part of any discrepancy. These data illustrate the problems and frustrations of estimating nitrogen cycling rates. Anyone attempting this task in a critical way performs a valuable service, not the least important aspect of which is to highlight areas of uncertainty.

The second attempt at a nitrogen budget for the Amazon area considers the river basin as a whole (Salati, Sylvester-Bradley & Victoria, 1982). Again the authors stress the assumptions made and the paucity of data. Comparing these data with those of Herrera & Jordan shows a number of similarities. First, most of the inorganic nitrogen entering the system with precipitation and leaving it by streamflow was ammonium: nitrate nitrogen was detectable, but always less than 8% of the concentration of ammonium. Second, organic nitrogen (particulate and soluble) was a feature of streamflow. Third, nitrogen fixation could be detected in nodulated legumes, associated with plant roots, in lichens and epiphytes (including cyanobacteria) and even in termites. As before, amounts varied in both space and time – this was especially true of the legume component. Fourth, input via precipitation and output via streamflow were approximately balanced. Thus, in order to balance the budget, gaseous losses must equal nitrogen fixation – estimated to be about 20 kg ha^{-1} y^{-1}. However, this time there are some other possibilities and unknowns. For example, in areas which have been cleared for agriculture, nitrogen may have been removed in crops. One further problem arises with nitrogen cycling in the humid tropics and that is that the rate of turnover may be so high that several cycles of breakdown, mineralization and uptake may occur in the one year: nitrogen budgets are usually drawn up on an annual basis.

Nitrification and nitrate in forest systems

Although nitrification was mentioned in the introductory part of this forest section, the examples given so far conform to the widely held opinion that in climax forests ammonium is the principal form of nitrogen and that by omitting the nitrification step, tight nitrogen cycling may be achieved and leaching of nitrate avoided. However, recent evidence from a wide variety of habitats indicates that this is not necessarily the case. Forests probably grow on soils covering a complete spectrum from those in which most of the nitrogen is in organic form, through ammonium as the major species to those in which nitrate plays a significant part. Unfortunately, with methods available until recently, it has not always been possible to distinguish between some of these options. Although expensive, use of the stable isotope ^{15}N can give valuable information on this point. Cleve & White (1980) added a small quantity of $K^{15}NO_3$ to a 60-y-old *Betula papyrifera* (paper birch) forest near Fairbanks, Alaska

(64°52'N, 147°02'W), and then looked at the proportion of different pools containing ^{15}N in the forest floor. Heal, Swift & Anderson (1982) recalculated some of these data and obtained figures indicating that most of the nitrogen was in the soluble organic fraction and that the turnover time of this fraction was much shorter and the daily flux much greater than for ammonium. Nitrate was present in only small amounts. Thus it seems likely that much of the nitrogen in this system is cycled through litter in the form of soluble organic compounds. On the other hand, a study of nine temperate forest systems in Madison, Wisconsin (43°04'N, 78°16'W), including five deciduous and four evergreen conifer sites, gives evidence of considerable nitrification and of nitrate uptake by trees (Nadelhoffer, Aber & Melillo, 1984). Indeed in only one case (*Pinus resinosa*) was less than 70% of the nitrogen taken up in the form of nitrate.

The actinorhizal genus *Alnus* sheds its leaves with an unusually high %N and these leaves are usually quickly broken down. At a high (1450 m) site near col d'Ornon, Isere (45°N, 06°E), Daniere, Capellano & Moiroud (1986) found rapid (<1 y) breakdown of *A. incana* leaves and production of up to 50 g nitrate-N ha^{-1} during the vegetative growth period. The soil had a high pH (7.4) and the nitrogen content of leaves only fell from 2.8% when they were fully green to 2.4% when they were dead. Like the alfisols of the Wisconsin study, the soils under *A. incana* were reasonably base saturated. It appears that, when extensive nitrification is found under forest trees, soils are relatively fertile. This does not mean that tree growth is nitrogen-sufficient, but that they can take up mitrate and also may respond better to fertilizer nitrogen than trees on more acid, nutrient-poor sites where ammonium is the major nitrogen form.

Peat soils

Peat is often thought of as a temperate or arctic phenomenon, arising when precipitation exceeds evaporation and decomposition rates are very slow. However, some peats are also found in the humid tropics.

Temperate bogs

As part of the International Biological Programme, an extensive study was carried out on the Moor House bog system in England (Heal & Perkins, 1978). Many characteristics of this site are more sub-arctic than temperate. The major features are listed in Table 5.3, and the nitrogen cycle illustrated in Figure 5.7. The pool of active nitrogen in the peat is around 100 kg h^{-1}, the rest of the 400–500 kg cycling is in the biomass

Table 5.3 *Some characteristics of Moor House bog, England (Heal &*
Perkins, 1978)

Peat-forming genera	*Sphagnum, Calluna, Eriophorum*
%N in peat	1.04–1.71
Ash (% dry weight)	3
C:N	~51
Peat depth (m)	2–3
Total N	20–30 t ha^{-1} of which 10% is actively cycled
pH	3.3–5.5
Redox potential (E_h)	500 mV at surface, down to 112 mV
Most prominent invertebrates	Diptera. (Few pollinators because few flowers)
Soil fauna	Rich: mites and assorted worms (but not earthworms). Saprovores
Herbivores, predators	Homopterans, red grouse, spiders, beetles, frogs, birds (meadow pipit)

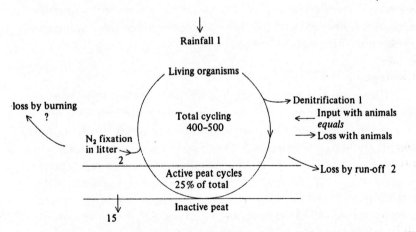

Figure 5.7. Nitrogen cycling in a temperature bog ecosystem (Heal & Perkins, 1978). All values are kg N ha^{-1} y^{-1}.

(plant, animal and microbial). There is very little ammonium and even less nitrate. Overall the site is thought to be climate, rather than nutrient limited.

The genera *Nitrobacter* and *Nitrosomonas* were not found, so that any nitrification must have been heterotrophic. However, most nitrate is probably brought in via precipitation. Organic nitrogen is brought to and taken from the area by animals; these two items seem to be approximately balanced. The bulk of the soluble nitrogen is organic (unlike the tundra system discussed earlier in which it was ammonium). Some of the soluble organic nitrogen passes down to the inactive peat layer, where it is immobilized; some is lost in run-off. Denitrification probably about balances nitrogen input via precipitation and there may be a small net input from heterotrophic nitrogen-fixing bacteria in litter. In the litter, fungi predominate over bacteria, but the reverse is true lower in the peat. Perhaps surprisingly in such a wet system, a significant part of the peat microflora is aerobic. Part of this is due to aerobic microsites forming around aerenchymatous plant tissues. Proteolytic bacteria are plentiful, but much of the ammonium released by them is rapidly immobilized. Natural or managed fires can result in considerable nitrogen loss – up to 50 kg ha^{-1} for a single burn. The overall picture is of a simple, tight nitrogen cycle.

An even simpler cycle obtains in a sub-arctic ombrotrophic mire in Sweden, where low pH prevents ammonia volatilization, nitrification and denitrification (Rosswall & Granhall, 1980).

Fens

These are also waterlogged, but are distinguished from bogs by having a supply of groundwater: bogs are entirely rain fed. Fens often have extensive peat deposits and many are more nutrient rich and of higher pH than bogs. One which has been widely studied is at Houghton Lake, Michigan (44°N, 84°W) (Richardson *et al.* 1978). The vegetation is principally composed of species of *Carex* and *Salix*. The top 20 cm of the peat contains most of the roots and the nutrients, there being little vertical water movement. The pH varies between 5.1 and 5.9. About 71% of the dry matter is organic (in bogs it is usually more than 95%) with nearly 30% ash, 2.54%N and 52 milliequivalents Ca^{2+} 100 g^{-1}. Figure 5.8 summarizes the annual nitrogen budget. Although the total nitrogen and the %N (1–2%) is higher than in many other bogs such as the Moor House bog, less of it is available – and so the biomass supported is much less (only plant material was included in the data for fen biomass, but even if there

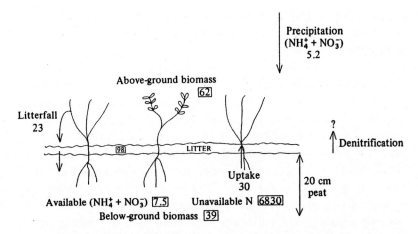

Figure 5.8. Annual nitrogen budget for a rich fen system. From Richardson *et al.* (1978). All values are in kg N ha^{-1}. Figures in rectangles are pool sizes, unenclosed figures indicate annual flows.

is an equal amount of animal + microbial biomass (very unlikely) this still holds true). Only one estimate was made of available (ammonium + nitrate) nitrogen, so too much weight should not be put on this figure. However, measurements of dissolved nitrogen in surface water and in interstitial water of the peat showed 52 and 142 mmol m^{-3} respectively of ammonium and 2.7 and 4 mmol m^{-3} of nitrate. Lower levels at the surface were thought to be due to algal and microbial uptake, but evidence of denitrification was also obtained. Plants took up 30, but only lost 23 kg N ha^{-1} each year. As a proportion of the whole of the available nitrogen (7 out of 26) this is large. Clearly if the plants had been growing in proportion to this rate of nitrogen accretion it would have been noticed! It is more likely that estimates were either inaccurate or did not include all components. For example there were no data on leaching or the effects of herbivores: these would account for a significant part of the discrepancy between losses from and gains to the biomass.

Tropical peats

These can cover quite large areas, possibly adding up to 20.5 million hectares in south-east Asia. Of these, 17 million hectares are in Indonesia, accounting for a possible 9% of that country's land surface; most of the rest are in Malaysia. As with temperate peats, they only occur when water accumulates, giving reduced soil oxygenation. In a preliminary consideration of nitrogen cycling in such a system Notohadiprawiro

(1981) stated that a further feature of these tropical peats is that they are oligotrophic. Detailed nitrogen-cycling studies on these interesting systems remain to be done.

Effects of flooding

Flooding may occur naturally when rivers burst their banks, or it may be used as a normal part of agricultural practice, for example in the culture of wetland rice. Studies on the latter may give a clue as to the likely effects of natural flooding on nitrogen cycling. Since wetland rice is the staple diet of much of Asia and elsewhere, in countries which are not rich, many studies have centred on the efficient use of both fertilizers and nitrogen-fixing plants such as *Azolla*, cyanobacteria and photosynthetic bacteria (Sprent, 1979; Skinner & Uomala, 1986). Careful timing of fertilizer application is necessary if major losses by denitrification are to be avoided (Bacon, McGarity, Hoult & Alter, 1986). For example, denitrification following nitrification may result in loss of over 50% of nitrogen applied as urea. When rice stubble is incorporated, nitrification is reduced because of the raised C:N ratio: ammonium is also immobilized. The net result is that denitrification losses are greatly reduced. Nitrification inhibitors and placement of fertilizers close to plant roots may also reduce losses (Keeney & Sahrawat, 1986; Reddy & Patrick, 1986). These various studies show how an understanding of nitrogen cycling can lead to adoption of management practices which can optimize nitrogen utilization by a crop.

6

Aquatic ecosystems

Introduction

All aquatic areas have some constraints on nitrogen cycling, the most common of which is that there will nearly always be an element of anaerobiosis. On the other hand, movement of ammonium is likely to be much freer than in soils because of the universal presence of water through which it can move and the general lack of cation-binding sites. Further, the animal:plant biomass ratio is higher (chapter 2). Lateral movement from one area to another, in addition to vertical movement is an extra complicating factor.

Gases diffuse in water about 10^5 times more slowly than in air. Thus it may take a considerable time for equilibria to be reached and gases in bulk fluid may be super- or under-saturated in some circumstances. The most frequent example of under-saturation is where oxygen is deficient in lower, unstirred water layers where it is used up by respiring organisms. On the other hand recent work has shown that both oxygen and nitrogen are frequently present at super-saturation in both open sea water (Craig & Hayward, 1987) and closed inland Antarctic lakes (Wharton, McKay, Mancinelli & Simmons, 1987). Oxygen super-saturation arises from a mixture of physical (barometric pressure fluctuations, downward movement of bubbles) and biological (photosynthetic) processes. In two stations in the North Pacific water at about 50 m depth reached, in summer, about 100% oxygen saturation (Craig & Hayward, 1987). Additional causes of super-saturation can be seen in ice-covered lakes, such as those on the Antarctic mainland. Here gases enter in solution in incoming water. When this freezes at the bottom of the ice cover, dissolved gases (including oxygen and nitrogen) are forced out into the remaining water. Nitrogen levels c. 150% saturation and oxygen levels considerably in excess of this (when active photosynthesis is occurring) may be found (Wharton et al., 1987). These factors have obvious implica-

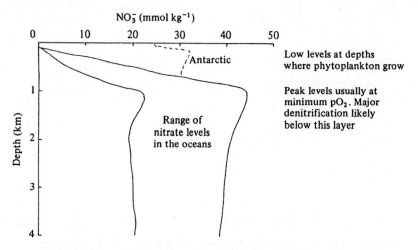

Figure 6.1. General pattern nitrate distributions in oceans, after Sharp (1983). Note that the surface pattern for Antarctic waters is distinct. Seasonal patterns are found in shallower waters when there are thermoclines followed by mixing.

tions for estimation of both net productivity (photosynthesis) and nitrogen cycle reactions.

The rapidity with which equilibrium is reached varies with many factors of which turbulence is one of the most important (as it is in air). A further difference between water and land-based systems lies in the depth of light penetration, which is much greater in water than in soil. Light may differentially affect microorganisms of the nitrogen cycle, apart from any effects on photosynthetic organisms. For example, nitrite-oxidizing species are more sensitive to light than ammonium-oxidizing species (Olson, 1980, quoted by Kaplan, 1983). These differences affect profiles of activity with depth as well as causing diurnal fluctuations. Light penetration will also affect the extent of photochemical reactions, of the type that occur in the atmosphere. These photochemical reactions can have implications for atmospheric nitrogen contents. For example, the dominant sources of nitric oxide in waters may come from photochemical dissociation of nitrite (Scranton, 1983): this nitric oxide may equilibrate with the atmosphere.

Biological denitrification may be the major process by which nitrogen gas in the atmosphere remains at about 79%. Much of this denitrification occurs at depth in the oceans, where oxygen is virtually unavailable and therefore dissimilatory nitrate reduction is at an advantage: the necessary nitrate is probably present here (Figure 6.1). Hattori (1983) gives values in excess of 350 Tg y^{-1}, most of which is in the eastern tropical north Pacific ocean. This compares with about 20 Tg of nitrogen gas fixed (Carpenter, 1983). The balance must be made from the input of combined nitrogen by run-off from the land.

Perhaps the simplest aquatic systems are more-or-less closed lakes and examples of these will be considered first.

Lakes

A special case: Lake Vanda, Antarctica

This lake has a number of unusual features (Canfield & Green (1985) and references therein) some of which are summarized in Figure 6.2. There is a permanent ice cover. Water, carrying low levels of salts, flows in seasonally from the Onyx river, and water may sublime from the surface. The net result is slowly increasing salinity. However, the main volume of the lake has relatively low salinity and circulating currents keep both it and temperature fairly constant. The water is very clear, but because of low nutrient availability (e.g. nitrate levels of about 5 mmol m^{-3}) there is little photosynthesis per unit volume. However, since the total stirred volume is so large, about half of the total lake photosynthesis takes place there. Peak photosynthesis per unit volume occurs in the virtually unstirred lower layer, close to the anoxic zone where phytoplankton adapted to low light is supported by the higher nutrient levels (nitrate about 100 mmol m^{-3}). All detected nitrification and nitrifying organisms were found in a narrow band beginning at the upper level of the layer where peak photosynthesis occurred. Maximum denitrification occurred just below the top of the anoxic zone. Note that the temperature at the lake bottom (68 m below the surface) is 25 °C. Because of the lack of circulating currents in the lower part, nutrient movement is diffusion limited. This dense, salty layer is thought to have been formed before the Onyx river began to flow and bring in less dense water. No mention of animal life was made, so we shall assume the nitrogen cycle proceeds via microbes and phytoplankton only. Nitrogenase activity was not detected at any depth (Vincent, Downes & Vincent, 1981). This very interesting study was slightly marred by the authors' use of the term 'denitrification' to cover both nitrate $\rightarrow$ nitrogen gas and nitrate $\rightarrow$ ammonium. Whilst they are not alone in this usage, it does lead to confusion. However, in spite of the uncertainties, it is possible to draw up a general nitrogen cycle

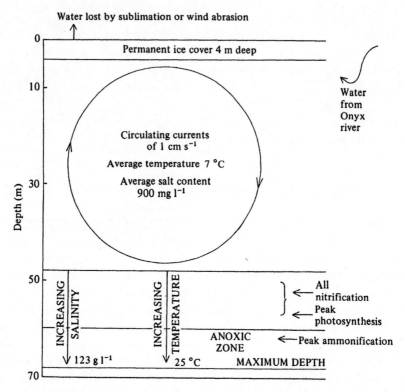

Figure 6.2. The major features of Lake Vanda, Antarctica. From Canfield & Green (1985).

for this area and this is done in Figure 6.3. Dead phytoplankton sinks to lower layers where it is broken down by microorganisms of which sulphate-reducing species are a significant component. Ammonium diffuses out of the anoxic zone and is converted to nitrate in the unstirred oxic layer (oxygen production by phytoplankton could be important here). Nitrate is used, either by the phytoplankton or it may diffuse downward and be reduced to nitrogen gas by denitrifying species. A possible variant of this is that N_2O is produced by nitrifying species and used by denitrifying species. The whole system is probably phosphorus, rather than nitrogen limited, and therefore little ammonium would become immobilized in the organic fraction. The vertical pattern of reactions is an unusual and interesting one and has been suggested as being a useful experimental tool (Vincent *et al.*, 1981).

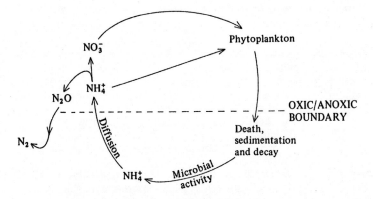

Figure 6.3. A general nitrogen cycle for the Lake Vanda ecosystem illustrated in Figure 6.2. From Canfield & Green (1985). Note that both ammonium and nitrous oxide diffuse across the oxic/anoxic boundary.

Freshwater lakes

Nitrogen cycling in most lakes is rather more complex than in Lake Vanda. Because of the concern about nitrate levels in drinking water and in pollution generally, the topic has received much attention in recent years. Table 6.1 attempts to summarize variables which affect the various reactions of the nitrogen cycle in deeper lakes where macrophytes are not a dominant component. These data are from temperate lakes where there are seasonal effects on water temperature gradients and concomitant effects on mixing. Superimposed on these variations are those due to the general level of nutrients, whether oligotrophic or eutrophic. Chief amongst these nutrients are nitrate entering from agricultural drainage water, and nitrate, organic nitrogen and phosphorus from sewage. The effects of increased nitrate loading depend on various factors. If other nutrients and environmental conditions are not limiting, primary productivity may be increased. If they are limiting, then both denitrification and dissimilatory nitrate reduction may be increased. Table 6.2 gives an example of this. Where nitrate levels are a problem, introduction of algae to immobilize nitrogen may be useful if other factors are not limiting. Otherwise it may be worthwhile to explore conditions for optimizing denitrification (Stewart, Preston, Peterson & Christofi, 1982). Encouragement of dissimilatory nitrate reduction does not reduce the nitrogen load. It is an irony of modern life that parts of the world need to enhance denitrification and other parts nitrogen fixation.

Table 6.1 *Factors affecting processes in the nitrogen cycle in deeper freshwater lakes without dominant macrophytes. Based on Stewart et al. (1982)*

Process	Factor/comment
Nitrogen fixation	Cyanobacteria usually at or near surface; photosynthetic bacteria at anoxic zone of shallow lakes. Chemoautotrophs and chemoheterotrophs may be important in sediments. Fixation at all levels only occurs when concentrations of soluble combined N are low
Mineralization	May occur in open water, but more important in sediments. Where plankton is the major biomass component mineralization likely to be rapid (low C:N)
Nitrification	Generally autotrophic and therefore dependent on both NH_4^+ and O_2. Site may vary with season, from sediment–water interface in spring and autumn to open water during summer (O_2 at interface depleted)
Denitrification	Activity seasonal, even when significant populations of appropriate bacteria are present. Affected by location and extent of NO_3^- supply. Generally occurs at sediment–water interface
Dissimilatory nitrate reduction	Found under similar conditions to denitrification, but essentially of much less significance to overall cycling
Nitrate assimilation	By phytoplankton. Varies with location and concentration of NO_3^-. Phytoplankton may also assimilate NH_4^+ and low molecular weight organic N compounds when available
Immobilization in living or dead biomass	Standing biomass varies greatly in quantity and position from season to season. N in sedimented dead organisms may be rapidly mineralized if sufficient O_2 available

Table 6.2 *Effects of nitrate load on some nitrogen cycle reactions in two freshwater lakes, Blelham (England) and Balgavies (Scotland). After Stewart et al. (1982). All figures in kg N ha^{-1}*

Component	Blelham	Balgavies
Nitrate load	46	507
Primary productivity	146	136
Denitrification	38	151
Dissimilation to ammonium	7	33

Marshes
Freshwater marshes

As soon as larger plants are involved, the C:N ratio of the biomass is increased and therefore the rate of mineralization is decreased: more nitrogen becomes fixed in biomass. This holds true for both marine and freshwater systems. Shallow edges of lakes may have macrophytes and thus a differently biased nitrogen cycle to the open water discussed above. Marshlands have a more uniform distribution of macrophytes. As an example, the freshwater segment of the Barataria Basin will be used. Figure 6.4 summarizes some of the properties of this area, which lies on the Mississippi delta, near New Orleans, above the north shore of the Gulf of Mexico (29°25′N, 89°50′W) (DeLaune, Smith & Sarafyan, 1986). The marsh is said to be flotant because its surface, which is dominated by the C_4 grass *Panicum hemitomon*, responds to increase in water level, i.e. it floats. In order to look for components of the nitrogen cycle and to test whether the site was nitrogen deficient, various doses of ammonium sulphate were given to circular plots. In some experiments, the ammonium was labelled with ^{15}N. Sedimentation was estimated by ^{137}Cs activity. This radio isotope has a long half-life and is virtually immobile in sediments. Accurate dating was possible because of fallout from early postwar nuclear testing (it's an ill wind . . .!). Additional nitrogen at 10 g m^{-2} gave a 40% increase in plant growth but 3 g m^{-2} gave none. This suggests that sufficient mineralization was occurring to support appreciable plant growth and 'swamp' the effect of small additions of nitrogen. Since ^{15}N was also given at 3 g m^{-2}, the fate of this isotope, when taken together with various nitrogen analyses, will reflect fairly accurately the processes of the unamended marsh. As can be seen from Figure 6.4, the budget does not balance! The figure of 120 kg ha^{-1} for the nitrogen in the 9-mm-deep band added to the peat layer each year is likely to be quite accurate, as is the 89 kg ha^{-1} in above-ground biomass. The value for nitrogen fixation is subject to error, being based on acetylene reduction assays over a long incubation time, but this error is unlikely to have underestimated by the shortfall found. If we assume that it is reasonably accurate, and noting that input via rainfall and loss by denitrification are small, then nitrogen must enter the system from elsewhere. This is thought to be from water, containing both ammonium and nitrate, flowing in from adjacent lakes. Low concentrations of ammonium were found throughout the 'soil': this fact, together with the high level of recovery of added nitrogen, is supporting evidence that mineralization occurs and that plants may absorb ammonium. The low level of denitrifi-

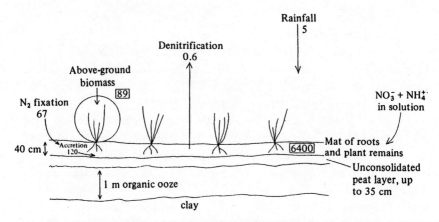

Figure 6.4. Nitrogen pools (figures in rectangles, kg ha^{-1}) and flows (kg ha^{-1} y^{-1}) for a freshwater marsh. The dominant macrophyte is *Panicum hemitomon*.

cation observed could be supported by nitrate in the incoming water. Alternatively, it could reflect the fact that some nitrification was occurring and that not all of the resultant nitrate was absorbed by plants. However, in a nitrogen-limited aquatic system such as this, little nitrification would be expected. In summary, this marsh represents an aggrading system where the additional nitrogen comes from both fixation by heterotrophs and by transfer from adjoining areas. The high accretion rate in terms of quantity and depth suggests that this stage in succession may not be long-lived. Marshes of this type could act as sinks for inorganic nitrogen in adjacent waters, in the same way as algae may act as a sink in the previous example. No data were available on nitrogen cycling within the plants.

Salt marshes

On the other side of the Florida peninsula to the Barataria Basin described above, at the estuary of the Altahama river (approximately 31°20′N, 81°30′W) in Georgia, is a salt marsh. In this area, Hopkinson & Schubauer (1984) studied the role of *Spartina alterniflora* in nitrogen cycling. In this marsh, salinity varies between 20 and 45 g per kg soil and the soil contains (by mass) 49% clay, 24% sand, 14% organic matter and 13% silt. The nitrogen content of various parts of the system was measured every 2 months for a year. Thus although microbial activity was not studied, a budget for flux through the plant part of the system allows some insight into overall nitrogen cycling (Figure 6.5). The major

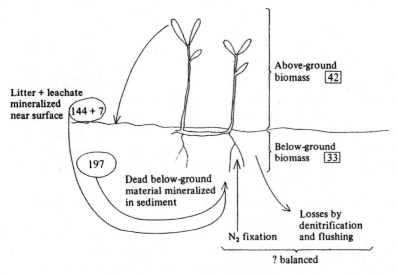

Figure 6.5. Nitrogen pools (figures in rectangles, kg ha^{-1}) and flows (in ovals, kg ha^{-1} y^{-1}) for a salt marsh dominated by *Spartina alterniflora*. From Hopkinson & Schubauer (1984).

simplifying assumption was that annual mortality equals annual production. This is consistent with the observation that the total nitrogen in living biomass remained about the same, although there was considerable seasonal variation in where the nitrogen was found. Rhizomes were a major store of both carbon and nitrogen in winter. The nitrogen in this store was used to drive leaf growth in spring. Rhizomes also acted as a route for the passage of nitrogen from senescent leaves of one shoot to the growing leaves of other shoots. Both the low %N (1.05 and 0.44 in above- and below-ground biomass respectively), and the high degree of internal recycling (estimated at 330 kg ha^{-1} y^{-1}) are indicative of nitrogen limitation. High rates of internal recycling also generally occur when there are high turnover rates (4.6 y^{-1} in this case). In discussing their data, Hopkinson & Schubauer (1984) quote evidence for mineralization in *Spartina* salt marshes and conclude that mineralized nitrogen is rapidly taken up by plants. They also note various reports on nitrogen fixation associated with *Spartina* (e.g. Patriquin & McClung, 1978); this process is known to occur only when the plants are actively photosynthesizing (Whiting, Gandy & Yoch, 1986). Losses occur by denitrification and by flushing. If the assumption of constant nitrogen in biomass is correct (i.e. the amount of nitrogen lost following death equals that taken up during growth), then nitrogen fixation must balance the sum of denitrification and loss by flushing. Once again, these processes can only be quantified very approximately, but in the range suggested (45–450 kg ha^{-1} y^{-1}) would represent a significant fraction of total nitrogen movement in the system.

Comparing their data with a freshwater *Scirpus* marsh, Hopkinson & Schubauer (1984) point out that *Spartina* has a lower %N and a higher level of nitrogen recycling within plants than *Scirpus*. They suggest that the *Scirpus* system might be more nitrogen sufficient than the *Spartina* system, possibly because the latter was in a saline environment. However, in the freshwater marsh described in the previous section the *Panicum* had only 0.67%N in above-ground biomass, even less than *Spartina*. *Panicum hemitomon* is a C$_4$ grass and may use its nitrogen (as well as its carbon dioxide) more efficiently than a C$_3$ plant (Brown, 1978). Clearly there are variations amongst macrophytes in freshwater as well as between freshwater and saline environments.

Other saline systems
Coral reefs
These only develop well in shallow (usually <10 m) warm (25–29 °C) waters, although older reefs, such as the Great Barrier Reef, may be much deeper. They are highly productive systems surrounded by seas which are very low in nutrients. Tight nutrient cycling is thus imperative. As far as nitrogen is concerned, this is achieved in a number of ways. Most of the corals, as well as other invertebrates, harbour symbiotic xooanthellae. These non-motile stages of dinoflagellates utilize ammonium excreted by the host coral and may also assimilate ammonium from outside the coral. Corals lacking such xooanthellae show net ammonium excretion (Burris, 1983, Figure 6.6). Thus within corals little combined nitrogen is likely to be lost. Cyanobacteria may fix nitrogen, generally only in the light, on the surface of coral reefs (e.g. Burris, 1976). These cyanobacteria are actively grazed by herbivores which excrete nitrogen into the surrounding medium. Sometimes they are also associated with nitrifying bacteria: the latter may be so dependent upon the cyanobacteria for ammonium that they appear to have light-dependent nitrate production. This nitrate may in turn be used by green algae (Webb & Wiebe, 1975).

For a developing coral reef there must be a net nitrogen input. This may come either from fixation or from efficient retrieval of ammonium or nitrate from local or more remote sources. Details must vary with site, but

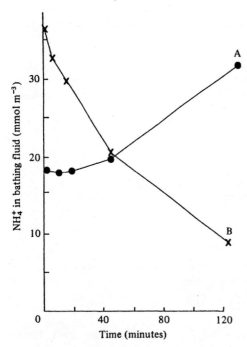

Figure 6.6. A, excretion of ammonium by coral *Dendophia nigrescens* which lacks xooanthellae. B, uptake of ammonium by coral *Seriatophora hystrix* which houses xooanthellae. Redrawn from Burris (1983).

photosynthetic organisms (green algae, dinoflagellates, cyanobacteria) are likely to play a major role in both nitrogen and carbon cycling.

Intertidal systems

These are subject to variations in salinity, oxygen supply, desiccation and other environmental factors. Even in unpolluted areas, considerable local variations in components of nitrogen cycling may be found. Various studies have attempted to estimate the activity of microorganisms of the nitrogen cycle. The study of Kaspar (1982, 1983) in New Zealand will be taken as an example, because several different sites within a comparatively small geographical area (3.1 km^2) near Nelson (approx. 41°10′S, 173°30′E) were used. These varied in flora and fauna as well as extent of tidal cover (Table 6.3). The marsh site had *Juncus maritimus* present and roots of this plant accounted for much of the 3.5–13.5% volatile solids (ones which are lost following treatment at 550 °C for 2 h) in the sediment: the sediment was mainly in reduced form, but oxidized areas were found along roots (roots of flooding-tolerant plants can often

Table 6.3 *Some features of estuarine sites on the Delaware inlet, near Nelson, New Zealand. From Kaspar (1983)*

	Site			
Factor	Marsh	Sand	Mud	Seagrass bed
Extent of flooding (hours per tidal cycle)	Spring high tides only	3–4	3–4	10–12
Denitrification from 0–10.5 cm depth ($mg N m^{-2} d^{-1}$)				
Mean *in situ* activity	4.08	2.94	4.14[a]	2.22
Mean *potential* activity	1943	163	281[a]	136
% recovery of nitrate-N in top 1.5 cm as				
N_2O	30.1	26.9	28.9	32.4
NH_4^+	5.2	11.4	7.6	6.9

[a] Estimated from yearly values (Kaspar, 1982)

transport oxygen from their shoot systems to their root systems, Armstrong & Wright, 1975). The solids in the mud site were also oxidized on the top and along walls of cavities made by mud snails (*Amphibola crenata*) and crabs (*Helice crassa*). Organic matter varied from 3.2% (upper layers) to 2.1% (lower layers). The sand site was also low in volatile organic matter, and was oxidized near the surface. It contained species of both *Euglena* and *Oscillatoria*, a nitrogen-fixing cyanobacterium also known as *Trichodesmium. Zostera novazelandica* was the major seagrass (and the major angiosperm) in the seagrass habitat, and below the oxidized surface layer the sediment in this site was heavily reduced. Denitrification was found at all sites, but the levels reached were only a very small proportion of the potential as indicated by incubation with added nitrate (Table 6.3). Dissimilatory nitrate reduction, yielding ammonium, varied between sites, being greatest in sand. Nitrate levels were usually lower than ammonium, reaching a maximum of about 5 mol m^{-3} in water on one occasion in the sand flat, but generally being much less than 1 mol m^{-3} and often below the detection limits. Although denitrification was found in all sites, the sand flats had a net nitrogen gain, because the amount of nitrogen fixed by *Oscillatoria* exceeded that lost by denitrification. These major differences over small areas again indicate the hazards of computing overall nitrogen budgets, but also suggest that management techniques could be developed to swing the balance in favour of denitrification or nitrogen fixation, depending on local needs and conditions.

Open sea

Open seas comprise approximately 70% of the surface of the world and over 99% of its inhabited space. Generally the nutrient content in open waters is low, but it increases towards shores. As discussed in chapter 2, nitrogen may be the factor limiting production in marine systems, whereas phosphorus is often limiting in fresh water. This may be related to molybdenum availability, since this element is required for both nitrogenase and nitrate reductase (Howarth & Cole, 1985). Nitrogen fixation may be carried out by both cyanobacteria and heterotrophic bacteria: many of the latter may be associated with phytoplankton. In view of recent evidence that *Azotobacter* can synthesize nitrogenase using vanadium instead of molybdenum (Bishop *et al.*, 1986) it is possible that some of these other, as yet largely unidentified bacteria (Guerinot & Colwell, 1985) may be able to fix nitrogen in an environment with very low levels of available molybdenum. Bacteria and cyanobacteria may be endosymbiotic with diatoms in the sea. Some of the symbioses are very fragile and their role may have been underestimated. For example nitrogenase activity may be detectable when the organisms are incubated in the sub-surface layers where they occur, but undetectable when they are lifted up to the deck of a ship for incubation. Martinez, Silver, King & Alldredge (1983) found aggregates of two *Rhizosolenia* spp. (filamentous diatoms) up to 7 cm in diameter, floating beneath the surface of various oceans. Within the cells of these diatoms were large numbers of bacteria, surrounded by membrane envelopes and capable, within their hosts, of high rates of nitrogenase activity. It was estimated that such bacteria enabled the *Rhizosolenia* to be self-sufficient in nitrogen. Local nitrogen fixation in deep oceanic hydrothermal vents has also been suggested (see references in Fogg, 1982).

Low levels of all forms of inorganic nitrogen, in addition to dissolved N_2 and other nitrogen-containing gases are present in sea water. Thus all the reactions of the nitrogen cycle are possible and likely. Inorganic nitrogen sources generally occur at very low concentrations, suggesting that microbial scavengers of dissolved nitrogen may be important. These include both bacteria (Jones & Rhodes-Roberts, 1980) and planktonic algae (Fogg, 1982 and references therein). Locally high concentrations of combined nitrogen (organic and inorganic) may occur, for example in the wake of excreting animals. A good ammonium or urea uptake system may

Table 6.4 *Some concentrations of nitrogen-containing substances present in six sites off Peru and California. Values from Williams (1967) as quoted by Liss (1975). Values are means in mmol m^{-3}, with ranges in parentheses*

Species	Surface 3 mm	Sub-surface layers from a minimum of 20 m to a maximum of 35 m
NH_4^+	11.1 (8.5–14.4)	0.9 (0.4–2.2)
NO_2^-	0.3 (0.0–1.0)	0.3 (0.1–1.2)
NO_3^-	2.7 (0.9–6.3)	4.5 (0.1–21.5)
Dissolved organic-N	20.7 (15.6–25.5)	5.2 (0.3–7.0)
Particulate organic-N	22.0 (11.7–60.0)	2.1 (0.4–3.7)

enable phytoplankton to make good use of these temporary periods of plenty.

In marine systems there is also great variation in content of nitrogen-containing substances with both location and depth. The surface 3 mm often has concentrations of particulate and dissolved substances which are quite different from those of the lower layers. Some representative values are given in Table 6.4. Nitrite does not seem to vary in a consistent way between surface and sub-surface layers, but the other components all tend to be more concentrated (up to 17 times) in the surface than in lower layers. This may be due to microbiological activity, although input from the atmosphere cannot be ruled out. Deeper in the oceans, nitrate nitrogen is generally much more concentrated than in the upper layers (Figure 6.1), where it is probably largely absorbed by phytoplankton (Spencer, 1975), even though many algae prefer ammonium. The increase in nitrate with depth is probably due to nitrification. Consideration of these variations, the number of samples which it is feasible to take, and the total surface and volume of oceanic waters, emphasizes the fact that estimates of global pools of nitrogen and of nitrogen cycling can only be approximate.

As with the Lake Vanda system discussed at the beginning of this chapter, particular nitrogen cycle reactions occurring at different depths are usually related to the oxygen profile and this in turn is related to light penetration and oxygen-evolving photosynthesis. Although not as extreme as on land, there are major variations in temperature throughout the world. This will affect solubility. For gases with low solubility, this could be important for nitrogen cycle reactions, but this aspect has been little considered.

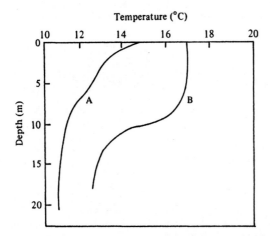

Figure 6.7. Temperature profiles of a frontal station (A) and a stratified station (B) near Vancouver Island. From Price, Cochlan & Harrison (1985).

Variations in water temperature profile are also found in coastal areas compared with deeper waters. This can have major effects. Figure 6.7 gives data from Price, Cochlan & Harrison (1985) for a study area in the Strait of Georgia in Western Canada. This strait runs between Vancouver Island and the mainland (48–50°N, 123–125°W). Two sites were studied, one in which the water was stratified and the other the frontal region between this and the completely mixed inshore waters. The frontal area had a high biomass of phytoplankton in the surface water accompanied by a high concentration (0.2–0.3 mmol m^{-3}) of nitrate relative to stratified water (<0.02 mmol m^{-3}). Phytoplankton species also varied considerably from predominantly diatoms in the frontal region to predominantly small flagellates in the open water. Larger numbers of zooplankton were found in the frontal than the stratified waters, and were thought to be an important factor in nitrogen cycling. As a result of experiments carried out during the summer season, using ^{15}N labelled nitrate, ammonium and urea (the phytoplankton used all of these and organic nitrogen, again stressing their versatility), it was concluded that distinct cycles operated, with the frontal areas losing organic nitrogen, probably to the stratified area, which appeared to take up more nitrogen than could otherwise be accounted for (Figure 6.8). In these studies no nitrogen-fixing organisms were detected.

Fogg (1982) summarized estimates available at the time for overall nitrogen cycling in the sea. He pointed out that total gains appear to be

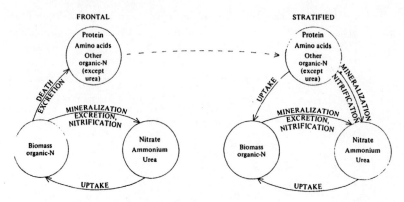

Figure 6.8. Nitrogen cycling in frontal and stratified zones of the Strait of Georgia. There is thought to be net loss from the frontal area and net input to the stratified area. From Price *et al.* (1985).

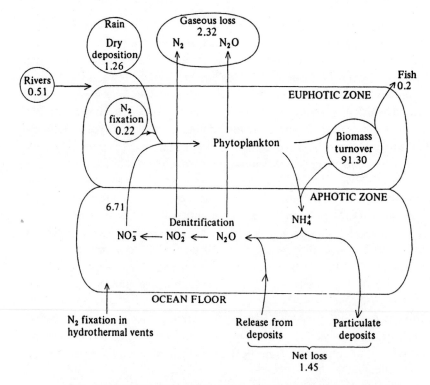

Figure 6.9. Possible nitrogen cycle for seas. Recalculated from Fogg (1982). Values given as percentages of nitrogen in the biomass, approximately 6900 Tg.

less than losses by about 124 Tg annually and that this situation could reflect either a real difference or inaccuracies in the estimates of some of the components (or both). Some of the difference could be offset by symbioses of the *Rhizosolenia* type, described above. Figure 6.9 summarizes the nitrogen cycle described by Fogg (1982), but expressing the data as a percentage of the nitrogen in the biomass (6900 Tg) for a particular year. Note that over 90% is in the organic fraction.

7

Impact of man

Introduction
With the exceptions of the use of fire and of mechanization, man and other animals do similar things to the environment (beavers fell trees, make dams; ants husband aphids, beetles farm fungi, etc.): man, however, does them to rapidly increasing extents. This chapter aims, not to denigrate man's effects on his environment, more to put them into the context of a potentially manageable nitrogen cycle.

Agriculture
Effects of cultivation
Many agricultural practices involve cultivation: this affects soil aeration and usually enhances mineralization of organic nitrogen in soil. Table 7.1 summarizes some effects found in a comparison of eight paired sites (virgin and cultivated) from seven states of the USA (Smith & Young, 1975). Once taken into cultivation, the total nitrogen and proportion of organic matter usually drop, although there are exceptions. Not surprisingly, the potentially mineralizable nitrogen also drops as does the carbon dioxide evolved (since microbial activity is reduced). In this particular case the rise in soil nitrate may be a combined effect of mineralization and residual fertilizer. We shall now examine some of these effects in more detail.

Use of fertilizers
Nitrogenous fertilizers are usually given as ammonium salts, anhydrous or liquid ammonia, nitrate or urea. All of these are rapidly incorporated into natural nitrogen cycling processes. For practical considerations, including cost of application and damage to standing crops, fertilizers tend to be given in one or a few heavy dressings per crop. Plants

Table 7.1 *Selected data from a comparison of eight paired soils[a], virgin versus cultivated. Data were for 0–15 cm depth and all differences were significant. From Smith & Young (1975)*

Factor	Virgin	Cultivated
%N	0.216	0.145[b]
% organic-C	2.60	1.51[b]
NO_3^- (mg kg^{-1})	7.3	12.5
Potentially mineralizable N (mg kg^{-1})	286	162
CO_2 production in soil (mg kg^{-1})	1460	722

[a] Soils were from Maine, Missouri (2), Minnesota, Texas, Ohio, New Mexico and Iowa. Virgin sites were either forest, native bush or grass. Cultivated plots were used for various crops for 25–75 y and given N fertilizer at rates from 13–336 kg ha^{-1} y^{-1}. Soil profiles down to 60 or 90 cm showed similar results, but of lower magnitude

[b] Two of the eight pairs (included in the means) were exceptions to the general rule that V > C

can only use a proportion of this immediately so problems such as leaching may easily arise (Table 7.2).

Nitrate is the most mobile form of nitrogen in most soils. It is also potentially toxic to humans, although opinions differ widely as to how big a threat this poses. There are two main possibilities, methaemoglobinaemia in young children and formation of carcinogenic nitrosamines in the human gut: the chemistry of these reactions is discussed in an excellent volume prepared by the National Academy of Sciences (Anon., 1978). There are agreed levels above which nitrate concentrations in drinking water should not rise – the World Health Organization recommends less than 11.3 mg N l^{-1}, although twice this limit is acceptable. This is equivalent to almost 1 mol m^{-3}, a level on which many plants thrive in flowing nutrient culture. Depending on the type of soil and underlying rock, nitrate may reach groundwater very quickly (weeks) or take many years if it has to percolate through rocks such as limestone. For example, it has been calculated that in the United Kingdom, nitrate moves downward through chalk and Triassic sandstones at 1 and 2 m y^{-1} respectively (Young, Oakes & Wilkinson, 1979; see also Wilkinson & Greene, 1982), This makes it possible to predict when problems in drinking water may arise (Foster, Cripps & Smith-Carington, 1982; Wilkinson & Greene, 1982). Studies of nitrate profiles in such systems also help to assess the extent of fertilizer loss, usually found to be in the range of 40–70% of

Table 7.2 *Possible fates of excess fertilizer nitrogen*

Form of N	Possible fate
Nitrate	Leaching into groundwater in wet areas. Conversion to N_2O and/or N_2 and loss to atmosphere. Occasionally bound in soil
Ammonium/ammonia	Often complexed with cation-binding sites in soil. May be volatilized and lost to atmosphere – especially at high pH
Urea	Usually rapidly hydrolysed in soil – then subject to same fates as ammonium/ammonia

fertilizer applied. This represents not only a potential pollution threat, but also a considerable waste of money.

Examination of soil and rock nitrate profiles can accurately reflect the recent history of land use. In particular, when permanent grassland is ploughed, mineralization occurs and the additional nitrate produces a front in the nitrate profile (Figure 7.1). The data show that permanent grassland can use most of the nitrogen applied in a moderate dose of fertilizer, giving few problems for groundwater. When such grassland is ploughed, there is a surge of mineralization, with a consequent rise in nitrate levels. Following this, growth of cereals with normal cultivation and moderate nitrogenous fertilization poses a real threat to groundwater. This partly results from the leaching of applied fertilizer and partly from the continued high rates of mineralization of residues after cultivation. These considerations, together with others, not the least of which is the desire to reduce costs, has led to the development of zero-tillage (also known as no-till) procedures, in which crop seed is sown directly into the residue of the previous crop. It is usual to use herbicides to control weeds. Zero-tillage practices also have their disadvantages, for example allelopathic effects and build-up of pest and pathogen populations, but offer considerable scope for minimizing nitrogen losses and improving the physical structure of soils. One of the most detailed studies to date on nitrogen cycling in tilled and no-tilled systems is that of House, Stinner, Crossley & Odum (1984). Nitrogen budgets were drawn up on the basis of large numbers of measurements of plant and animal numbers and biomass. It was concluded that no-till systems gradually (in time) diverge from conventional tillage systems and that they may tend towards a nitrogen cycle typical of natural terrestrial ecosystems. This should mean that nitrate pollution problems can be minimized.

Various ways of removing nitrate from groundwater have been consi-

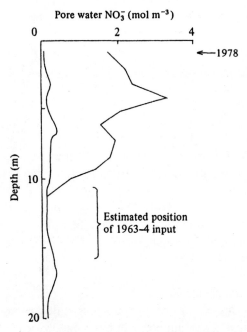

Figure 7.1. Nitrate profiles for a chalk soil in East Yorkshire (England). The left-hand profile was for permanent pasture which had been given 676 kg fertilizer-N in the previous 12 y. The right-hand profile was from part of the same land which had been ploughed and used for winter or spring sown cereals for 12 y from 1966, during which time it had received at least 815 kg N ha^{-1} (figures were unavailable for 1976). Profile dating was based on a tritium reference. From Foster *et al.* (1982).

dered (Anon., 1978; Wilkinson & Greene, 1982). Some of these are very simple: one which can be used for surface run-off is to collect it in reservoirs before it is used for piped water supplies: the remedy is time! If allowed to stand for a period (6 months is often appropriate) then denitrification may reduce the nitrate concentration to safe levels. However, management problems can arise when the major nitrate influxes occur at the time of peak water demand. A logical extension of this use of natural denitrification is to enhance it, which usually means adding a suitable carbon substrate for the denitrifying organisms (see also Table 7.7 and the section on use of wetlands, p. 128).

Slow-release fertilizers

It would obviously be desirable to produce fertilizers in such a way that nitrogen is only slowly released into the soil. Table 7.3 lists some of the

Table 7.3 *Some types of slow-release nitrogen-containing fertilizers*

Organic formulation	E.g. substituted ureas – examples isobutylidene, diurea
Coated inorganic pellets	E.g. sulphur-coated ureas, plastic-coated granules
Blended pellets	Mixtures of substances with differing solubilities

methods which have tried to do this. In all cases the product is more costly to manufacture than the more usual forms of nitrogenous fertilizer, making them at present a practical proposition only for specialist purposes (turf on sports grounds, high-value crops) or on soils where leaching or denitrification are particularly bad problems. With proper management (suitable size of fertilizer particle, suitable application time with respect to irrigation) substituted ureas, for example, may have a role in wetland rice production (Rubio & Hauck, 1986).

Correct timing and use of foliar fertilizers
Careful timing of fertilizer application can often effect major savings. In particular, autumn applications of ammonia in temperate agriculture are often very wasteful because nitrification proceeds over winter, and when moist soils warm in spring both leaching of nitrate and denitrification can occur. Use of foliar fertilizers potentially ensures that the nitrogen reaches the plant rather than the microorganisms of the soil. In some cases it is very effective (Embleton *et al.*, 1986), but there are many technical and economic problems still to be solved.

Use of inhibitors
Urease inhibitors
Much of the nitrogenous fertilizer used today is in the form of urea. Although plants can take up urea, they usually do not have the opportunity, because of the widespread occurrence of urease as a soil enzyme (Burns, 1978). In many areas ammonia produced by the hydrolysis of urea may be volatilized (see p. 59). A specific inhibitor of urease would prevent this problem and an active search for an effective and safe inhibitor has been prosecuted for some years. Since the problem is particularly acute in alkali soils (because of the effect of high pH on ammonia volatilization), an example from such an area will be taken. One of the recent success stories in land reclamation has been that of the alkali soils around the Central Soil Salinity Research Institute in Karnal, India. Rao & Ghai (1986) studied the effects of four urease inhibitors on

Table 7.4 *Effects of the urease inhibitor phenylphosphorodiamidate on apparent nitrogen uptake efficiency by wheat grown in alkali soil with different levels of urea nitrogen. Efficiency measured as*

$$\frac{N \ uptake \ in \ treated \ plants \ - \ N \ uptake \ in \ unfertilized \ plants}{N \ applied} \times 100$$

Urea N given (mg kg^{-1})	Efficiency	
	Inhibitor absent	Inhibitor present
40	53.0	70.6
80	36.9	39.6
120	27.0	35.2

nitrogen uptake and growth of wheat in a greenhouse experiment using local soil of a pH of 9.3 (the wheat was selected to grow at this pH) and a soil urease activity of 12.5 μg urea hydrolysed per g soil per hour. Table 7.4 gives the nitrogen uptake by wheat in the presence and absence of the most successful urease inhibitor, phenylphosphorodiamidate. A significant improvement in the efficiency of nitrogen uptake by plants was found, but, as the authors stress, the costs of inhibitor application must be offset against the benefits of increased nitrogen uptake before such procedures are adopted by farmers. In some situations, nitrogen given in the form of urea may be lost at a later stage, after both ammonification and nitrification, as a result of denitrification (Duxbury & McConnaughey, 1986).

Nitrification inhibitors
Whereas ammonia losses are a major problem in some areas, leaching of nitrate (or its denitrification) is the major problem in others. One way of reducing this loss is to use nitrification inhibitors and this has been successful in many areas. Table 7.5 lists some of the commercially available nitrification inhibitors. One of the most widely studied and used is N-serve, which selectively inhibits *Nitrosomonas*. N-serve is volatile and difficult to incoporate into soil, adding application costs to production costs: some farmers therefore prefer to use excess fertilizer to balance estimated losses (Briggs, 1975). However, there has been some success in giving combined dressings of N-serve and ammonia. In some trials, use of nitrification inhibitors has resulted in increased nitrogen uptake by crops, in other cases the effects are more subtle. For example, Juma & Paul (1983) used ATC on wheat plants grown in the field in Canada; ^{15}N

Table 7.5 *Some commercially available nitrification inhibitors.*
From Hauck (1983) and Juma & Paul (1983)

Trade or trivial name	Compound
N-serve, nitrapyrin	2-chloro-6(trichloromethyl)pyridine
'AM'	2-amino-4-chloro-6-methyl pyrimidine
ATC	4-amino-1,2,4-triazole
MAST	2-mercapto-benzothiazole
'ST'	sulfathiazole
dicyandiamide	dicyandiamide
thiourea	thiourea

labelled urea or NH_4OH were used as fertilizers, with and without ATC. There were no significant differences in ^{15}N in the wheat grain or straw. However, significantly more ^{15}N was retained in the soil of the ATC-treated plots. Further analyses of the soil component (which included roots) revealed that there was twice as much ^{15}N in the microbial biomass of the ATC-treated soils and it was concluded that this nitrogen would be slowly released over a period of time. Therefore, in this case the effects of the nitrification inhibitor were long-term rather than immediate. There is likely to be a whole spectrum of responses, depending on the nature of the particular soil and its environment (Keeney, 1986). Possible consequences of inhibition of nitrate formation include accumulation of fairly readily available ammonium, and diversion of this ammonium to other processes, including immobilization and volatilization.

Use of legumes and other nitrogen-fixing plants

Legumes have been used since pre-Roman times, with the intention of improving soil fertility as well as providing crops. In terms of minimizing pollution, they are useful because they fix nitrogen in relation to the plant's needs, reducing leaching problems. However, on death of any part of the plant, nitrogen will be mineralized and the possibilities for leaching will be the same as for any plant residue. Ploughing in of legume residues may therefore have several effects. First, they increase the content of nitrogen in the soil, possibly leading to a surge in nitrate in leachate (as described above). Second, if more nitrogen is removed with the crop than was fixed, there will be a net *loss* of soil nitrogen. Residues are then likely to have a high C:N ratio and soil nitrogen may be immobilized (Sprent, 1986b). Direct transfer of nitrogen from legumes to associated plants is an important consideration, but one which has proved to be extraordinarily difficult to quantify. Transfer may occur after mineralization of senescent legume parts (but remember the nitrogen is equally available to the legume if it is still growing). Direct transfer has recently been discounted in the field in the short-term (36 days) (Ledgard, Freney & Simpson, 1985). Use of legumes in intensive (rather than low-input – low-output) systems requires considerable management skill, but is now receiving renewed attention (e.g. Patriquin, 1986).

Azolla has been used in agriculture, particularly in rice cultivation. For an account of the use of *Azolla* and other nitrogen-fixing organisms in agriculture see Moore (1969) and the volume edited by Skinner & Uomala (1986). Detailed budgets of their role in the nitrogen cycle have not been drawn up.

Use of fallow periods

In many semi-arid areas, bare fallow periods (one where no crop is grown) are included in the crop rotation: during this time the water in the soil is re-charged – farmers in such areas sometimes claim to farm for water, not for yield. This procedure is widely used in the wheat-growing prairies of the USA and Canada. During the fallow periods, organic residues are mineralized and soil nitrate levels may increase greatly, but because of the dry climate are not usually leached. A recent study (Lamb, Peterson & Genster, 1985) shows how tillage system may affect this mineralization. The study area was in the High Plains area (*c.* 1600 m) of Nebraska (41°N, 103°W), where winter wheat is grown. Three fallow treatments were used: (a) ploughing in spring of the fallow year (bare-fallow); (b) no-till, in which weed control was by a contact herbicide; (c) a stubble mulch system in which stubble from the previous crop was mulched using a sweep plough. Fallow and wheat treatments alternated and the experiment was continued for 12 years. Two sites with different soil characteristics were used. Major site-to-site and year-to-year differences in the accumulation of soil nitrate were found. In the first three years, more nitrate was present at both sites following ploughing than in the other two treatments. Subsequently, this effect of ploughing was lost, probably because bacterial populations became stabilized. Seasonal profiles of soil nitrate nitrogen for this stable period, averaged over the three tillage treatments, are shown in Figure 7.2. Note the rapid increase during the hot, dry summer period. For management purposes it is important to understand seasonal variation in a given area in order to advise on fertilizer requirements. In this example, advice based on a June sample could have resulted in the use of excess nitrogenous fertilizer. This would not only be expensive, but also lead to crop management problems

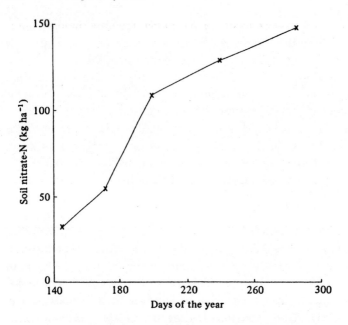

Figure 7.2. Seasonal accumulation of nitrate during a fallow year in Nebraska (USA). After Lamb *et al.* (1985).

such as leaching and lodging (collapse of stems under wind or rain because too much nitrogen makes the tissues soft).

Organic nitrogen application

Table 7.6 gives the nitrogen content of various animal residues. It has been practice for many centuries to use animal (including human) waste on soils, in addition to that which is returned naturally during grazing. This was once a matter of good husbandry, but now, with intensive animal rearing techniques, it is also a waste disposal problem. Whereas droppings from battery hens are easily dried, slurry from housed cattle and pigs is 90% water. This is heavy and expensive to transport, to say nothing of the offensive odour. Nitrogen is readily lost as ammonia and organic compounds: the extent of this process varies with temperature. Much research is now being conducted into the optimum use of slurries, including minimizing pollution problems. It has been estimated that a pig-fattening unit of 10 000 places has the pollution equivalent of municipal sewage from 18 000 inhabitants! Nitrogen cycle reactions in such slurries have been extensively discussed (e.g. Anon., 1978). There may be various consequences, for example accumulation of nitrite

Table 7.6 *Average nitrogen in dry matter of various manures and slurries. From Gostick (1982)*

Source	%N	Possible problems for plant growth
Farmyard manure – cattle	2.4	
– pig	2.4	
Poultry – broiler litter	3.4	
– battery hen droppings	6.0	excess P, Ca
Cattle slurry	5.0	excess K
Pig slurry	6.0	excess P, Cu

because high pH prevents its oxidation (Figure 3.6). Recent increases in the extent of fish farming have raised the possibility of locally high levels of excreted and excess food-derived nitrogen compounds in both fresh and inshore marine waters. Organic nitrogen and ammonium can find their way from animal slurries and municipal waste into rivers and other waterways. The former leads to growth of heterotrophic bacteria which reduce the oxygen concentration and result in the depletion of fish stocks.

Disposal of sewage nitrogen

Most industrial nations with large human populations have sewage treatment plants. The technology of these was originally aimed at removing suspended solids and organic matter which would encourage the growth of oxygen-consuming microorganisms, rather than controlling soluble nitrogen. Attention is now being focussed on the nitrogen problem: one way of controlling this is to incorporate into treatment plants selected stages of the biological nitrogen cycle (see Table 7.7), using one of several possible designs of treatment plant. Alternatives to this are to use chemical methods. One such is the addition of lime to raise the pH and help to volatilize ammonia; ammonia is then stripped from the finely sprayed water by a large volume of air. This literally dilutes the problem – a concentrated ammonia solution becomes an atmosphere with low levels of ammonia. The residual waste from this process must be reacidified before release. At the moment, even with well-designed treatment systems, much nitrogen remains in the sludge and this itself presents problems. Some alternatives for coping with this are listed in Table 7.8.

This part of the chapter has so far concentrated on problems which developed countries have in coping with their (and their animals')

Table 7.7 *Sequence of biological nitrogen cycle reactions used to remove nitrogen from sewage or animal effluent*

Process	Comment
Ammonification of organic-N	Uses most of the available C. Autotrophic nitrifiers will not grow well until available C is low
$NH_4^+ \rightarrow NO_3^-$	A necessary prerequisite for denitrification! Normal autotrophic nitrifiers (*Nitrosomonas, Nitrobacter*) used. May be low-temperature limited in winter in some areas. Requires O_2
$NO_3^- \rightarrow N_2$	Requires addition of a C source (methanol often used) for most denitrifiers. Use of S-oxidizing denitrifier, such as *Thiobacillus denitrificans*, a possibility. Requires anoxic conditions. If N_2O to be final product pH must be >7, if less, N_2O formed

Table 7.8 *Some methods of sludge dispersal in use in the USA*

Chemical processing	Pyrolysis, wet oxidation, incineration
Potentially biological processes	Composting, land-fill, liquid or dry application to land, dispersal into lagoons or oceans

effluent. However, in many underdeveloped countries the priorities are quite different. Paradoxically, many of the warmer parts of the world such as India and most semi-arid parts of Africa (both of which may also have cold winters) have an acute shortage of fuel for cooking. This has resulted in extensive use of cow manure as fuel. Cows, being sacred to the Hindu religion, flourish in cities as well as the country and dung can be seen stacked for drying by the roadside and carried as a valued possession.

Use of natural and artificial wetlands

Use of plants of various types to absorb solutes such as nitrogen compounds (or other pollutants including heavy metals), may help to alleviate waste disposal problems. In the case of algae, they have the added benefit of restoring dissolved oxygen. Many algae prefer ammonium to nitrate and thus nitrification may be minimized (and parallel to this, the oxygen demand reduced). For a model of the possible effects of algae on waste entering the Great Miami river during its passage from Ohio to Fairfield see Bingham, Lin & Hoag (1984). In general algae

Table 7.9 *Some macrophytes with potential for wastewater nitrogen treatment. See Stowell, Ludwig, Colt & Tchobanoglous (1981); O'Brien (1981); Wolverton, McDonald & Duffer (1983)*

Category of plant	Examples	Comments
Floating	*Eichornia* spp. (water hyacinth)	Not cold tolerant; root system has very high surface to volume ratio, but not very deep (200 mm). High rate of transpiration could drop water level
	Ludwigia spp. (water primrose)	Deeper roots (600 mm), but poorer root surface properties for bacterial colonization
	Lemna, Spirodela, Wolffia spp. (duckweeds)	Some are cold tolerant; easy to harvest; shallow rooting; rapid growth gives continuous water cores which may restrict O_2 exchange but suppresses growth of mosquito larvae
Emergent	Various monocotyledonous genera including *Juncus, Scirpus, Typha* (reeds, rushes, cattails)	Many are cold tolerant, submerged stems throughout water column, roots in sediment; depth limited; permit good O_2 exchange between water and air; can be adapted for hydroponics and for use with fixed bacterial-support surface such as rocks and shores
Submerged	Various, including algae and higher plants	Possibly less easily managed; marked diurnal effects on O_2 in water, acting as source (day) and sink (night)

are not popular for use in control of nitrogen pollution for various reasons including the odour which they emit after they die.

Ideally, if plants are to be used in a practical way they must be harvested and preferably used for an economic purpose such as green manure. So far this ideal has not generally been reached. Indeed the major role of macrophytes in water treatment systems may be to provide a surface for microbial action! It is in this role that they are being developed in artificial wetlands, designed for the treatment of municipal wastewater. Such systems may be economically competitive for small to medium-sized communities (Gersberg, Elkins & Goldmann, 1984), provided of course that land is available and the climate suitable. Some possible plants are listed in Table 7.9. The actual choice will depend on climate, water depth and many other factors including other pollutants such as heavy metals (which some plants ad- or absorb strongly). Although nitrogen scavenging by plants is considered only as a minor role, its extent probably varies with species and conditions. Duckweeds, for example, contain more reduced nitrogen (about 4.6% of dry weight) than water hyacinth (2.9%) grown under similar conditions (O'Brien, 1981), probably because of their high proportion of photosynthetic to non-photosynthetic tissues. Species almost certainly vary in their preferences for ammonium, nitrate or other forms of soluble nitrogen and this will affect how they compete with microorganisms of the nitrogen cycle.

Although the potential for these systems is now widely accepted, much remains unknown as to how they operate biologically, in particular how nitrogen is lost from the aqueous to the gaseous phases (Handley-Raven, pers. comm.). The major possibilities are NH_3, NO_x and N_2. Of these N_2 is the most desirable, but its production will depend very much on factors such as substrate, species of microorganism, pH and temperature.

All of these problems are encountered in growing of wetland rice. Here, however, the emphasis is on maximizing the uptake of soluble nitrogen compounds by the plant and minimizing losses by volatilization and denitrification. Recent reviews have highlighted such systems and suggested possible management practices to overcome these problems (Bacon *et al.*, 1986; Keeney & Sahrawat, 1986; Reddy & Patrick, 1986; see also p. 100).

Forestry

For many years utilization of trees for timber or fuel was largely a matter of felling what was needed. Nowadays in many parts of the world there is a move towards reafforestation. This has meant much more interest in and research into the effects of felling on nutrient cycling and the nutrient requirements for replanting, as well as the likelihood of leached nitrogen reaching surface waters.

The effects of felling on nitrogen cycling

Most of the nitrogen in mature forests is recycled; small losses are replaced by one or more of many small inputs (e.g. from lichens, heterotrophic nitrogen-fixing species in litter, nitrogen contained in rainwater). Felling and clearing of trees may remove up to 700 kg N ha^{-1} (Heal *et al.*, 1982). Residues may be burnt, leading to further loss, and finally soil processes will alter if uptake by tree roots no longer occurs. These latter aspects have been studied in detail by Vitousek *et al.* (1982)

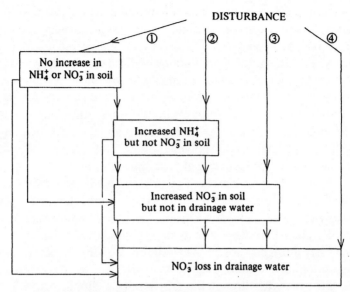

Figure 7.3. Diagram to show possible sequences of nitrogen-cycling events following disturbance in forests. The numbers refer to the classes of effect described by Vitousek *et al.* (1982) and discussed in the text.

for 17 different forests (including both deciduous and evergreen species) in the USA at sites ranging from the north-west and north-east to New Mexico. In each forest, plots were cleared and areas within the clearings trenched and lined with plastic so that soil processes could be monitored. The major difference between the experimental system and most managed systems was that weeds were not allowed to grow. Four types of response were observed, sometimes in sequence at a single site (Figure 7.3). In the first, no change of ammonium concentration was observed. There are two possible reasons for this (a) no significant net mineralization took place; (b) mineralization = volatilization and/or immobilization. Volatilization was ruled out on the grounds of low soil pH and on results from experiments with ^{15}N labelled substrate. Immobilization varied between sites, but was never great. It was concluded that this type of response occurred in sites where the trees prior to felling were nitrogen limited, giving litter which was slow to decompose because of its high C : N ratio. In the class 2 type of response, mineralization occurs but nitrification does not. Again the latter could be absence of the process (nitrification) or nitrification = denitrification and/or nitrate immobilization. A lag in nitrification was thought to be the most likely cause. Of the various reasons for such a lag, two were identified as being most likely, both requiring that nitrogen concentrations in the original forest soil were low. First, if trees had competed successfully with nitrifying microorganisms for the soil nitrogen, then populations of these organisms would be low. This is probable as the affinity of nitrifying organisms for ammonium is known to be low. Thus, after felling, there would be a lag during which populations of nitrifying organisms built up and nitrate became detectable. Second, that plants release polyphenols under conditions of nutrient limitation and that these polyphenols inhibit growth of nitrifying organisms. Such suppression could continue after felling.

The third and fourth classes of response allow for significant nitrification and only differ in whether or not the nitrate generated is leached. This will depend on how much nitrate is produced and how much water is available. These two classes occur where overall nutrient status is higher, so that more nitrogen is lost with litterfall and resorption of nitrogen from foliage prior to abscission may be low. In other words, comparatively more nitrogen cycles through the soil, as opposed to the plant.

These findings have implications for both nitrate pollution of water and for re-vegetation, either natural or after planting. Vegetation renewal will be more rapid and less fertilizer will be required for replanting on high-nutrient sites (see also chapter 4).

Reafforestation

Clearly this may require a considerable nitrogen input, which may be achieved by use of either fertilizer or nitrogen-fixing plants, as in agriculture. In the latter case, actinorhizal species are frequently more prominent than legumes, although this may be in part because most work has been done in temperate systems (Gordon & Wheeler, 1983). Problems may arise if the growth of the nitrogen-fixing species cannot be kept in check during the establishment phase. However, this can usually be overcome by management and selection procedures. It may be feasible to harvest both nitrogen-fixing and associated species and this possibility is particularly attractive in short-term rotations. Studies on nitrogen cycling in such systems are in progress, for example in Quebec, Canada, where alder (*Alnus glutinosa*) and hybrid poplar (*Populus nigra* × *P. trichocarpa*) are being grown. In this area, growth of the usually aggressive alder may be restricted by water supply, but it nevertheless fixed 40–60 kg N ha^{-1} y^{-1} – not a world record figure, but good for the unproductive site used (Côté & Camiré, 1985).

There is, however, a further possibility for reafforestation in peaty soils, namely release of nitrogen from the considerable stores present in the peat. Planting on peat is usually preceded by digging of drainage channels and the drying of peat is further encouraged by the uptake and transpiration of water by young trees. Conditions thus become more aerobic and conducive to mineralization. The detailed effects depend strongly on peatland type and content of nutrient other than nitrogen (Williams, Cooper & Pyatt, 1979).

Debris from trees
Naturally, as a result of wind-felling, or artificially, following logging (especially where timber is transported by water), wood debris is often found in forest waterways. For the Pacific NW area of the USA, Aumen, Bottomley & Gregory (1985) suggest that such debris might act as a sink for exogenous nitrogen which in turn would stimulate wood decomposition and then activate other nitrogen cycling processes.

Fire
Fire can occur naturally or as a result of accident, and the results can be severe. In areas prone to wild fire, such as those with a mediterranean climate, including large parts of Australia, fire is used as a managment practice, as it has been on a smaller scale by native peoples including the Australian Aborigines for thousand of years. Controlled burning removes forest litter and can be manipulated so that trees are not killed but can resprout (in the case of eucalypts and many other Australian plants) from buds deep inside the wood. These buds, together with seedlings from species with heat-stimulated germination, rapidly restore a leaf canopy. The effect of managed fire is to damp the oscillations caused by intermittent wild fires. However, decisions on the frequency of fires needed to avoid loss of endangered species and to support agriculture and forestry requirements may be difficult to make (Bell, Hopkins & Pate, 1982; see also Stock & Lewis, 1986). Although nitrogen-fixing plants frequently grow rapidly after fire, recent measurements suggest that, at least in dry areas, they may not fix enough nitrogen to replace that lost (Hansen, Pate, Hansen & Bell, 1987). In soils of low nutrient status, there may be a steady decline in nitrogen under a managed fire system (Lawrie, 1986).

As well as the obvious effects of fire on the above-ground biomass and removal of litter, there will also be effects on soil. Deposition of ash and

Table 7.10 *Effects of fire on soil processes related to nitrogen cycling. After Raison (1979)*

Factor	Possible effects	Comments
Ash	Increased N-cycling rates. Increased N_2 fixation. pH changes leading to release of substances from soil organic matter	Effects vary with nature of ash, soil and climate
	Increased nitrification	Little direct evidence
Heat	Loss of N	High-temperature decomposition of organic matter
	Increase in N content of residual matter	May occur up to 500 °C. Associated reduction in C:N ratio could accelerate N cycling
	Increase in ammonium	Possibly a combined effect of release from soil minerals and physico-chemical release from soil organic matter. Starts at about 100 °C, peaks at 250–300 °C
	Decrease in nitrate	Nitrates decompose above 150 °C
	Differential killing of organisms may increase soil ammonium	Nitrifiers generally killed $\leqslant$58 °C; some spore-forming ammonifiers can survive 100 °C

the nature of that ash will depend on the severity of the fire and on wind conditions. These factors, together with the nature of the soil itself, will affect the temperature profile achieved in the soil and this in turn will affect the processes influenced by fire. Some of these are summarized in Table 7.10, see also p. 92; they are further discussed in Raison (1979).

Acid rain and related phenomena

When fossil fuels are burnt, by coal or oil-fired power stations, by domestic fires, by internal combustion engines or by any other means, substances are vented into the atmosphere which may return to the surface in an acid form, in solution, or as dry deposition: these substances are collectively known as acid rain (Kennedy, 1986). Acid rain consists approximately of 70% sulphur and 30% nitrogen compounds, giving rise to sulphuric and nitric acids respectively. The chemistry of production of these is far from simple and has recently been reviewed (Calvert *et al.*, 1985). We are concerned only with the effects these substances may have on the nitrogen cycle: they fall into two types. First, acidification of the environment: this has so far been most widely studied in fresh waters. Although potentially alarming in proportion, in principle the effects of pH on nitrogen cycle processes are as described earlier (chapter 3). Second, effects on plant life, particularly trees. Because these are a major nitrogen-containing component of the biosphere, any forests killed by acid rain will result in major perturbations of the nitrogen cycle. However, the extent to which forest damage is caused by acid rain *per se* or by other pollutants is by no means clear (Anon., 1986). Acid absorption by leaves may be rendered harmless if the leaves have a sufficient buffering capacity. Recent evidence suggests that the buffering capacity of leaves may be insufficient to cope with the high levels of acidic gases in the atmosphere which are found in industrial areas, such as parts of Germany (Pfanz & Heber, 1986). Death of trees and other plants probably results from a combination of several atmospheric pollutants, including acids. On a more positive side, plants can assimilate low levels of nitrogen from acid rain, but this is thought to be insignificant in productivity terms, at least for crop plants (Evans *et al.*, 1986).

Third, there is a direct effect of burning of fossil fuels on global nitrogen pools. Although the total nitrogen will not be affected, the proportion of that total in a biologically active form will be greatly increased. Depending on the type of combustion, the product may be reduced nitrogen (mainly ammonia) or nitrogen oxides. Up to 100 Tg N per year in these forms may be emitted to the atmosphere. Because much of it comes from point sources (for example, power stations) local concentrations may be very high. Direct oxidation of nitrogen gas also occurs in high-temperature combustions and, again, local concentrations may be high. Although in global terms the proportion of this attributable to jet aircraft is low, because emissions are located near the tropopause, effects are potentially very great on the whole structure of the atmosphere (see chapter 2 and Hahn & Crutzen, 1982).

References

Adams, M. A. & Attiwill, P. M. (1986a). Nutrient cycling and nitrogen mineralization in eucalypt forests of south-eastern Australia. I. Nutrient cycling and nitrogen turnover. *Plant and Soil*, **92**, 319–39.

Adams, M. A. & Attiwill, P. M. (1986b). Nutrient cycling and nitrogen mineralization in eucalypt forests of south-eastern Australia. II. Indices of nitrogen mineralization. *Plant and Soil*, **92**, 341–62.

Alexander, I. J. (1983). The significance of ectomyorrhizas in the nitrogen cycle. In *Nitrogen as an Ecological Factor, 22nd Symposium of the British Ecological Society*, ed. J. A. Lee, S. McNeill and I. H. Rorison, pp. 69–93. Oxford: Blackwell Scientific Publications.

Allen, O. N. & Allen, E. K. (1981). *The Leguminosae – A Source Book of Characteristics, Uses and Nodulation*. London: Macmillan.

Anderson, J. M., Leonard, M. A., Ineson, P. & Huish, S. (1985). Faunal biomass: a key component of a general model of nitrogen mineralization. *Soil Biology and Biochemistry*, **17**, 735–7.

Anderson, O. E., Boswell, F. C. & Harrison, R. M. (1971). Variation in low temperature adaptability of nitrifiers in acid soils. *Soil Science Society of America Proceedings*, **35**, 68–71.

Andrews, M. (1986). The partitioning of nitrate assimilation between root and shoot of higher plants. *Plant, Cell and Environment*, **9**, 511–19.

Anon. (1978). *Nitrates: An Environmental Assessment*. Washington DC: National Academy of Sciences.

Anon. (1983). *The Nitrogen Cycle of the United Kingdom. A Study Group Report*. London: The Royal Society.

Anon. (1986). *Report of the Acid Rain Enquiry*. Edinburgh: The Scottish Wildlife Trust.

Armstrong, W. & Wright, E. J. (1975). Radial oxygen loss from roots: the theoretical basis for the manipulation of flux data obtained by the cylindrical platinum electrode technique. *Physiologia Plantarum*, **35**, 21–6.

Asghar, M. & Kanehiro, Y. (1976). Effect of sugarcane trash and apple residue incorporation on soil nitrogen, pH and redox potential. *Plant and Soil*, **44**, 209–18.

Atkinson, M. J. & Smith, S. V. (1983). C:N:P ratios of benthic plants. *Limnology and Oceanography*, **28**, 568–74.

Aumen, N. G., Bottomley, P. J. & Gregory, S. V. (1985). Nitrogen dynamics in stream wood samples incubated with [^{14}C] lignocellulose and potassium [^{15}N] nitrate. *Applied and Environmental Microbiology*, **49**, 1119–23.

Bacon, P. E., McGarity, J. W., Hoult, E. H. & Alter, D. (1986). Soil mineral nitrogen concentration within cycles of flood irrigation: Effect of rice stubble and fertilization management. *Soil Biology and Biochemistry*, **18**, 173–8.

Bavel, C. H. M. van & Baker, J. M. (1985). Water transfer by plants from wet to dry soil, *Naturwissenschaften*, **72**, 606–7.

Bell, D. T., Hopkins, A. J. M. & Pate, J. S. (1982). Fire in the Kwongan. In *Kwongan – Plant Life of the Sand Plain* ed. J. S. Pate & J. S. Beard, pp. 178–204. Nedlands: University of Western Australia Press.

Bingham, D. R.. Lin, C-H. & Hoag, S. (1984). Nitrogen cycle and algal growth modeling. *Journal of the Water Pollution Control Federation*, **56**, 1118–22.

Bishop, P. E., Premakumar, R., Dean, D. R., Jacobson, M. R., Chisnell, J. R., Rizzo, T. M. & Kopczynski, J. (1986). Nitrogen fixation by *Azotobacter vinelandii* strains having deletions in structural genes for nitrogenase. *Science*, **232**, 92–4.

Bliss, L. C., Heal, O. W. & Moore, J. J. (eds.) (1981). *Tundra Ecosystems: A Comparative Analysis*. The International Biological Programme 25. Cambridge: Cambridge University Press.

Bloom, A. J., Chaplin, F. S. & Mooney, H. A. (1985). Resource limitation in plants – an economic analogy. *Annual Review of Ecology and Systematics*, **16**, 363–92.

Bolin, B. & Cook, R. B. (eds.) (1983). *The Major Biogeochemical Cycles and their Interactions*. Scope 21. Chichester: John Wiley & Sons.

Boudot, J. P. & Chone, Th. (1985). Internal nitrogen cycling in two humic-rich acidic soils. *Soil Biology and Biochemistry*, **17**, 135–42.

Bowen, G. D. & Smith, S. E. (1981). Effect of mycorrhizas on nitrogen uptake by plants. In *Terrestrial Nitrogen Cycles*, Ecological Bulletin 33, ed. F. E. Clark & T. Rosswall, pp. 237–48. Stockholm: Swedish Natural Science Research Council.

Briggs, G. C. (1975). The behaviour of the nitrification inhibitor 'N-Serve' in broadcast and incorporated applications to soil. *Journal of the Science of Food and Agriculture*, **26**, 1083–92.

Brown, R. H. (1978). A difference in N use efficiency in C_3 and C_4 plants and its implications in adaptation and evolution. *Crop Science*, **18**, 93–8.

Bunnell, F. L., Maclean, S. F. Jr & Brown, J. (1975). Barrow, Alaska, USA. In *Structure and Functions of Tundra Ecosystems*, Ecological Bulletin 20, ed. T. Rosswall & O. W. Heal, pp. 73–124. Stockholm: Swedish Natural Science Research Council.

Burns, R. G. (ed.) (1978). *Soil Enzymes*. London: Academic Press.

Burns, T. A. Jr, Bishop, P. E. & Israel, D. W. (1981). Enhanced nodulation of leguminous plant roots by mixed cultures of *Azotobacter vinelandii* and *Rhizobium*. *Plant and Soil*, **62**, 399–412.

Burris, R. H. (1976). Nitrogen fixation by blue–green algae of the Lizard Island area of the Great Barrier Reef. *Australian Journal of Plant Physiology*, **3**, 41–51.

Burris, R. H. (1983). Uptake and assimilation of $^{15}NH_4^+$ by a variety of corals. *Marine Biology*, **75**, 151–5.

Calvert, J. G., Lazarus, A., Kok, G. K., Heikes, B. G., Walega, J. G., Lind, J. & Cantrell, C. A. (1985). Chemical mechanisms of acid generation in the troposphere. *Nature*, **317**, 27–35.

Campbell, R. (1985). *Plant Microbiology*. London: Edward Arnold.

Canfield, D. E. & Greem. W. J. (1985). The cycling of nutrients in a closed-basin antarctic lake: Lake Vanda. *Biogeochemistry*, **1**, 233–56.

Canuto, V. M., Levine, J. S., Augustsson, T. R. & Imhoff, C. L. (1982). UV radiation from the young sun and oxygen and ozone levels in the prebiological atmosphere. *Nature*, **296**, 816–20.

Carpenter, E. J. (1983). Nitrogen fixation by marine *Oscillatoria* (*Trichodesium*) in the world's oceans. In *Nitrogen in the Marine Environment*, ed. E. J. Carpenter & D. G. Capone, pp. 65–103. New York: Academic Press.

Chalk, P. M. & Smith, C. J. (1983). Chemodenitrification. In *Gaseous Loss of Nitrogen from Plant–Soil Systems*. Developments in Plant and Soil Sciences, vol. 9, ed. J. R. Freney & J. R. Simpson, pp. 65–89. The Hague. Martinus Nijhoff/Dr W. Junk.

Charley, J. L. & McGarity, J. W. (1964). High soil nitrate levels in patterned saltbush communities. *Nature*, **201**, 1351–2.

Clarholm, M. (1985). Interactions of bacteria, protozoa and plants leading to mineralization of soil nitrogen. *Soil Biology and Biochemistry*, **17**, 181–7.

Cleve, K. van & Alexander, V. (1981). Nitrogen cycling in tundra and boreal ecosystems. In *Terrestrial Nitrogen Cycles*, Ecological Bulletin 33, ed. F. E. Clark & T. Rosswall, pp. 375–404. Stockholm: Swedish Natural Science Research Council.

Cleve, K. van & White, R. (1980). Forest-floor nitrogen dynamics in a 60-year-old paper birch ecosystem in interior Alaska. *Plant and Soil*, **54**, 359–81.

Côté, B. & Camiré, C. (1985). Nitrogen cycling in dense plantings of hybrid poplar and black alder. *Plant and Soil*, **87**, 195–208.

Craig, H. & Hayward, T. (1987). Oxygen supersaturation in the ocean: biological versus physical contributions. *Science*, **235**, 199–201.

Daft, M. J. & El-Ghiami, A. A. (1976). Studies on nodulated and mycorrhizal peanuts. *Annals of Applied Biology*, **83**, 273–6.

Daniere, C., Capellano, A. & Moiroud, A. (1986). Dynamique de l'azote dans peuplement natural d'*Alnus incana* (L.) Moench. *Acta Oecologica/Oecologia Plantarum*, **7**, 165–75.

Davis, E. A. & Debano, L. F. (1986). Nitrate increases in soil water following conversion of chaparral to grass. *Biogeochemistry*, **2**, 53–65.

DeLaune, R. D., Smith, C. J. & Sarafyan, M. N. (1986). Nitrogen cycling in a freshwater marsh of *Panicum hermitomon* on the deltaic plain of the Mississippi river. *Journal of Ecology*, **74**, 249–56.

Doerge, T. A., Bottomley, P. J. & Gardner, E. H. (1985). Molybdenum limitations to alfalfa growth and nitrogen content on a moderately acid high-phosphorus soil. *Agronomy Journal*, **77**, 895–901.

Dommergues, Y., Garcia, J-L. & Ganry, F. (1980). Microbiological considerations of the nitrogen cycle in west African ecosystems. In *Nitrogen Cycling in West African Ecosystems*, ed. T. Rosswall, pp. 55–72. Uppsala: SCOPE/UNEP.

Duxbury, J. M. & McConnaughey, P. K. (1986). Effect of fertilizer source on denitrification and nitrous oxide emissions in a maize-field. *Soil Science Society of America Journal*, **50**, 644–8.

Embleton, T. W., Matsumura, M., Stolzy, L. H., Devitt, D. A., Jones, W. W., El-Motaium, R. & Summers, L. L. (1986). Citrus nitrogen fertilizer management, groundwater pollution, soil salinity, and nitrogen balance. *Applied Agricultural Research*, **1**, 57–64.

Eskew, D. L., Welch, R. M. & Cary, E. E. (1983). Nickel: an essential nutrient for legumes and possibly all higher plants. *Science*, **222**, 621–3.

Evans, L. S., Canada, D. C. & Santucci, K. A. (1986). Foliar uptake of ^{15}N from rain. *Environmental and Experimental Botany*, **26**, 143–6.

Fahey, T. J., Yavitt, J. B., Pearson, J. A. & Knight, D. H. (1985). The nitrogen cycle in lodgepole pine forests, southeastern Wyoming. *Biogeochemistry*, **1**, 257–75.

Faria, S. M. de, Lima, H. C. de, Franco, A. A., Mucci, E. S. F. & Sprent, J. L. (1987). Nodulation of legume trees from south east Brazil. *Plant and Soil*, **99**, 347–56.

Fay, P. (1981). Photosynthetic micro-organisms. In *Nitrogen Fixation, Volume 1: Ecology*, ed. W. J. Broughton, pp. 1–29. Oxford: Clarendon Press.

Fillery, I. R. P. (1983). Biological denitrification. In *Gaseous Loss of Nitrogen from Plant–Soil Systems*. Developments in Plant and Soil Sciences, vol. 9, ed. J. R. Freney & J. R. Simpson, pp. 33–64. The Hague: Martinus Nijhoff/Dr W. Junk.

Firestone, M. K. (1982). Biological denitrification. *Agronomy*, **22**, 289–336.

Focht, D. D. (1974). The effect of temperature, pH, and aeration on the production of nitrous oxide and gaseous nitrogen – a zero-order kinetic model. *Soil Science*, **118**, 173–9.

Focht, D. D. & Verstraete, W. (1977). Biochemical ecology of nitrification and denitrification. *Advances in Microbial Ecology*, **1**, 135–214.

Fogg, G. E. (1982). Nitrogen cycling in sea waters. *Philosophical Transactions of the Royal Society London, B*, **296**, 511–20.

Foster, S. S. D., Cripps, A. C. & Smith-Carington, A. (1982). Nitrate leaching to groundwater. *Philosophical Transactions of the Royal Sciety London, B*, **296**, 477–89.

Francis, R. & Read, D. J. (1984). Direct transfer of carbon between plants connected by vesicular-arbuscular mycorrhiza mycelium. *Nature*, **307**, 53–6.

Franco, A. A. & Munns, D. W. (1981). Responses of *Phaseolus vulgaris* L. to molybdenum under acid conditions. *Soil Science Society of America Journal*, **45**, 1144–8.

Freckman. D. W. & Whitford, W. G. (1987). The role of soil biota in decomposition in the Chihuahuan desert. In *Arid Lands: Today and Tomorrow*, ed. G. Nabhan & A. Elias-Cesnik. Tucson: University of Arizona, in press.

Freney, J. R. & Simpson, J. R. (1983). *Gaseous Loss of Nitrogen from Plant–Soil Systems*. Developments in Plant and Soil Sciences, vol. 9. The Hague: Martinus Nijhoff/Dr W. Junk.

Galbally, I. E. & Roy, C. R. (1983). The fate of nitrogen compounds in the atmosphere. In *Gaseous Loss of Nitrogen from Plant–Soil Systems*. Developments in Plant and Soil Sciences, vol. 9, ed. J. R. Freney & J. R. Simpson, pp. 265–84. The Hague: Martinus Nijhoff/Dr W. Junk.

Gersberg, R. M., Elkins, B. V. & Goldmann, C. R. (1984). Wastewater treatment by artificial wetlands. *Water Science and Technology*, **17**, 443–50.

Gordon, J. C. & Wheeler, C. T. (eds.) (1983). *Biological Nitrogen Fixation in Forest Ecosystems: Foundations and Applications*. The Hague: Martinus Nijhoff/Dr W. Junk.

Gostick, K. G. (1982). Agricultural development and advisory service (ADAS) recommendation to farmers on manure disposal and recycling. *Philosophical Transactions of the Royal Society London, B*, **296**, 329–32.

Graham, T. W. G., Webb, A. A. & Waring, S. A. (1981). Soil nitrogen status and pasture productivity after clearing of brigalow (*Acacia harpophylla*). *Australian Journal of Experimental Agriculture and Animal Husbandry*, **21**, 109–18.

Grant, M. A. & Hochstein, L. I. (1984). A dissimilatory nitrite reductase in *Paracoccus halodenitrificans*. *Archives of Microbiology*, **137**, 79–84.

Griffin, D. M. (1981). Water and microbial stress. *Advances in Microbial Ecology*, **5**, 91–136.

Grove, A. T. (1985). The arid environment. In *Plants for Arid Lands*, ed. G. E. Wickens, J. R. Goodin & D. V. Field, pp. 9–18. London: George Allen & Unwin.

Guerinot, M. L. & Colwell, R. R. (1985). Enumeration, isolation, and characterization of N_2-fixing bacteria from seawater. *Applied and Environmental Microbiology*, **50**, 350–5.

Hahn, J. & Crutzen, P. J. (1982). The role of fixed nitrogen in atmospheric photochemistry. *Philosophical Transactions of the Royal Society London, B*, **296**, 521–41.

Hansen, A. P., Pate, J. S., Hansen, A. & Bell, D. T. (1987). Nitrogen economy of post-fire stands of shrub legumes in Jarrah (*Eucalyptus maginata* Donn ex Sm.) forests of S.W. Australia. *Journal of Experimental Botany*, **38**, 26–41.

Harris, R. F. (1982). Energetics of nitrogen transformations. In *Nitrogen in Agricultural Soils*. Agronomy Monograph no 22, ed. F. J. Stevenson, pp. 833–89. Madison, WI: ASA-CSSA-SSSA.

Hart, M. H. (1979). Was the pre-atmosphere of the earth heavily reducing? *Origins of Life*, **9**, 261–6.

Hattori, A. (1983). Denitrification and dissimilatory nitrate reduction. In *Nitrogen in the Marine Environment*, ed. E. J. Carpenter & D. G. Capone, pp. 191–232. New York: Academic Press.

Hauck, R. D. (1983). Agronomic and technological approaches to minimizing gaseous nitrogen losses from croplands. In *Gaseous Loss of Nitrogen from Plant–Soil Systems*. Developments in Plant and Soil Sciences, vol 9, ed. J. R. Freney and J. R. Simpson, pp. 285–312. The Hague: Martinus Nijhoff/Dr W. Junk.

Heal, O. W. & Perkins, D. F. (ed.) (1978). *Production Ecology of British Moors and Montane Grassland*. Ecological Studies 27. Berlin: Springer-Verlag.

Heal, O. W., Swift, M. J. & Anderson, J. M. (1982). Nitrogen cycling in United Kingdom forests: the relevance of basic ecological research. *Philosophical Transactions of the Royal Society London, B*, **296**, 427–44.

Herrera, R. & Jordan, C. F. (1981). Nitrogen cycle in a tropical Amazonian rain forest: the caatinga of low mineral nutrient status. In *Terrestrial Nitrogen Cycles*, Ecological Bulletin 33, ed. F. E. Clark & T. Rosswall, pp. 493–505. Stockholm: Swedish Natural Science Research Council.

Hopkinson, C. S. & Schubauer, J. P. (1984). Static and dynamic aspects of nitrogen cycling in the salt marsh graminoid *Spartina alterniflora*. *Ecology*, **65**, 961–9.

House, G. J., Stinner, B. R., Crossley, D. A. Jr & Odum, E. P. (1984). Nitrogen cycling in conventional and no-tillage agro-ecosystems: analysis of pathways and processes. *Journal of Applied Ecology*, **21**, 991–1012.

Howarth, R. W. & Cole, J. J. (1985). Molybdenum availability, nitrogen limitation, and phytoplankton growth in natural waters. *Science*, **229**, 653–5.

Huang, T-C. & Chow, T-J. (1986). New type of N_2-fixing unicellular cyanobacterium (blue–green alga). *FEMS Microbiology Letters*, **36**, 109–10.

Ingham, E. R., Trofymow, J. A., Ames, R. N., Hunt, H. W., Morley, C. R., Moore, C. J. & Coleman, D. C. (1986a). Trophic interactions and nitrogen cycling in a semi-arid grassland soil. I. Seasonal dynamics of the natural populations, their interactions and effects of nitrogen cycling. *Journal of Applied Ecology*, **23**, 597–614.

Ingham, E. R., Trofymow, J. A., Ames, R. N., Hunt, H. W., Morley, C. R., Moore, C. J. & Coleman, D. C. (1986b). Trophic interactions and nitrogen cycling in a semi-arid grassland soil. II. System responses to removal of different groups of soil microbes or fauna. *Journal of Applied Ecology*, **23**, 615–30.

Jeter, R. M. & Ingraham, J. L. (1981). The denitrifying prokaryotes. In *The Prokaryotes*, ed. M. P. Starr, H. Stolp, H. G. Trüpes, A. Balows & H. G. Schlegel, pp. 913–25. Berlin: Springer-Verlag.

Jones, K. & Bangs, D. (1985). Nitrogen fixation by free-living heterotrophic bacteria in an oak forest: the effect of liming. *Soil Biology and Biochemistry*, **17**, 705–9.

Jones, K. G. C. (1985). Nitrogen fixation as a control in the nitrogen cycle. *Journal of Theoretical Biology*, **112**, 315–32.

Jones, K. L. & Rhodes-Roberts, M. E. (1980). Physiological properties of nitrogen-scavenging bacteria from the marine environment. *Journal of Applied Bacteriology*, **49**, 421–33.

Juma, N. G. & Paul, E. A. (1983). Effect of a nitrification inhibitor on N immobilization and release of ^{15}N from nonexchangeable ammonium and microbial biomass. *Canadian Journal of Soil Science*, **63**, 167–75.

Kaplan, W. A. (1983). Nitrification. In *Nitrogen in the Marine Environment*, ed. E. J. Carpenter & D. G. Capone, pp. 139–90. New York: Academic Press.

Kaspar, H. F. (1982). Denitrification in marine sediment: Measurement of capacity and estimate of *in situ* rate. *Applied and Environmental Microbiology*, **43**, 522–7.

Kaspar, H. F. (1983). Denitrification, nitrate reduction to ammonium, and inorganic nitrogen pools in intertidal sediments. *Marine Biology*, **74**, 133–9.

Kawama, A. (1981). Nitrogen cycling in tropical forests. In *Nitrogen Cycling in S.E. Asian Wet Monsoonal Ecosystems*, ed. R. Wetselaar, J. R. Simpson & T. Rosswall, pp. 119–22. Camberra: Australian Academy of Science.

Keeney, D. R. (1986). Inhibition of nitrification in soils. In *Nitrification*, Special Publications of the Society for General Microbiology, vol. 20, ed. J. I. Prosser. pp. 99–115. Oxford: IRL Press.

Keeney, D. R. & Sahrawat, K. L. (1986). Nitrogen transformation in flooded rice soils. *Fertilizer Research*, **9**, 15–38.

Kennedy, I. R. (1986). *Acid Soil and Acid Rain*. Chichester: John Wiley & Sons.

Kessel, J. F. van (1978). Gas production in aquatic sediments in the presence and absence of nitrate. *Water Research*, **12**, 291–7.

Kinjo, T. & Pratt, P. F. (1971). Nitrate adsorption: I. In some acid soils of Mexico and South America. *Soil Science Society of America Proceedings*, **31**, 722–5.

Knowles, R. (1982). Denitrification. *Microbial Reviews*, **46**, 43–70.

Knowlton, S. & Dawson, J. O. (1983). Effects of *Pseudomonas cepacia* and cultural factors on the nodulation of *Alnus rubra* roots by *Frankia*. *Canadian Journal of Botany*, **61**, 2877–82.

Lamb, R. J. (1985). Litter fall and nutrient turnover in two eucalypt woodlands. *Australian Journal of Botany*, **33**, 1–14.

Lamb, J. A., Peterson, G. A. & Fenster, C. R. (1985). Fallow nitrate accumulation in a wheat-fallow rotation as affected by tillage system. *Soil Science Society of America Journal*, **49**, 1441–6.

Lamont, B. (1982). Mechanisms for enhancing nutrient uptake in plants with particular reference to Mediterranean South Africa and Western Australia. *Botanical Reviews*, **48**, 597–689.

Larson, R. J. (1986). Water content, organic content, and carbon and nitrogen composition of medusae from the northeast Pacific. *Journal of Experimental Marine Biology and Ecology*, **99**, 107–20.

Lawrie, A. C. (1986). Nitrogen fixation by woody plants in natural ecosystems. In *Proceedings of the Eighth Australian Nitrogen Fixing Conference*. Australian Institute of Agricultural Science, Publication Number 25, ed. W. Wallace & S. E. Smith, pp. 97–9. Melbourne: Australian Institute of Agricultural Science.

Ledgard, S. F., Freney, J. R. & Simpson, J. R. (1985). Assessing nitrogen transfer from legumes to associated grasses. *Soil Biology and Biochemistry*, **17**, 575–7.

Lee, K. E. & Wood, T. G. (1971). *Termites and Soils*. London: Academic Press.

Levine, J. S. & Augustsson, T. R. (1985). The photochemistry of biogenic gases in the early and present atmosphere. *Origins of Life*, **15**, 299–318.

Lindeboom, H. J. (1984). The nitrogen pathway in a penguin rookery. *Ecology*, **65**, 269–77.

Lindstrom, K., Sarsa, M-L., Polkunen, J. & Kansanen, P. (1985). Symbiotic nitrogen fixation of *Rhizobium* (*Galega*) in acid soils. *Plant and Soil*, **87**, 293–302.

Lipsett, J. & Simpson, J. R. (1973). Analysis of the response by wheat to application of molybdenum in relation to nitrogen status. *Australian Journal of Experimental Agriculture and Animal Husbandry*, **13**, 563–6.

Liss, P. S. (1975). Chemistry of the sea surface microlayer. In *Chemical Oceanography*, vol. 2, 2nd edn, ed. J. P. Riley & G. Skirrow, pp. 193–244. London: Academic Press.

Lloyd, D., Boddy, L. & Davies, J. P. (1987). Persistence of bacterial denitrification capacity under aerobic conditions: the rule rather than the exception. *FEMS Microbiology Ecology*, **45**, 185–90.

Lombin, G. (1985). Evaluation of the micronutrient fertility of Nigeria's semi-arid savannah soils: boron and molybdenum. *Soil Science and Plant Nutrition*, **31**, 13–25.

Ludwig, R. A. (1984). *Rhizobium* free-living nitrogen fixation occurs in specialised non-growing cells. *Proceedings of the National Academy of Sciences, USA*, **81**, 1566–9.

Mancinelli, R. L. & Hochstein, L. I. (1986). The occurrence of denitrification in extremely halophilic bacteria. *FEMS Microbiology Letters*, **35**, 55–8.

Marion, G. M., Miller, P. C., Kummerow, J. & Oechel, W. C. (1982). Competition for nitrogen in a tussock tundra ecosystem. *Plant and Soil*, **66**, 317–27.

McCarty, G. W. & Bremner, J. M. (1986). Effects of phenolic compounds on nitrification in soil. *Soil Science Society of America Journal*, **50**, 920–3.

McClung, G. & Frankenberger, W. T. Jr (1985). Soil nitrogen transformations as affected by salinity. *Soil Science*, **139**, 405–11.

McComb, A. J., Cambridge, M. L., Kirkman, H. & Kuo, J. (1981). The biology of Australian seagrasses. In *The Biology of Australian Plants*, ed. J. S. Pate & A. J. McComb, pp. 251–93. Nedlands: University of Western Australia Press.

McEwan, A. G., Greenfield, A. J., Wetzstein, H. G., Jackson, J. B. & Ferguson, S. J. (1985). Nitrous oxide reduction by members of the family Rhodospirillaceae and the nitrous oxide reductase of *Rhodopseudomonas capsulata*. *Journal of Bacteriology*, **164**, 823–30.

McGill, W. B., Hunt, H. W., Woodmansee, R. G. & Reuss, J. O. (1981). Phoenix, a model of the dynamics of carbon and nitrogen in grassland soils. In *Terrestrial Nitrogen Cycles*, Ecological Bulletin 33, ed. F. E. Clark and T. Rosswall, pp. 49-115. Stockholm: Swedish Natural Science Research Council.

Martinez, L., Silver, M. W., King, J. M. & Alldredge, A. L. (1983). Nitrogen fixation by floating diatom mats: a source of new nitrogen to oligotrophic ocean water. *Science*, **221**, 152–4.

Medina, E. (1982). Nitrogen balance in the Trachypogon grasslands of central Venezuela. *Plant and Soil*, **67**, 305–14.

Monteith, J. & Webb, C. (eds.) (1981). *Soil Water and Nitrogen in Mediterranean-Type Environments*. The Hague: Martinus Nijhoff/Dr W. Junk.

Moore, A. W. (1969). *Azolla*; Biology and agronomic significance. *Botanical Reviews*, **35**, 17–34.

Myrold, D. D. & Tiedje, J. M. (1986). Simultaneous estimation of several nitrogen cycle rates using ^{15}N: Theory and application. *Soil Biology and Biochemistry*, **18**, 559–68.

Nadelhoffer, K. J., Aber, J. D. & Melillo, H. M. (1984). Seasonal patterns of ammonium and nitrate uptake in nine temperate forest ecosystems. *Plant and Soil*, **80**, 321–35.

Nagata, T. (1986). Carbon and nitrogen content of natural planktonic bacteria. *Applied and Environmental Microbiology*, **52**, 28–32.

Neilands, J. B. & Leong, S. A. (1986). Siderophores in relation to plant growth and disease. *Annual Review of Plant Physiology*, **37**, 187–208.

Niell, F. X. (1976). C:N ratio in some marine macrophytes and its possible ecological significance. *Botanica Marina*, **19**, 347–50.

Nohrstedt, H-O. (1985). Nonsymbiotic nitrogen fixation in the topsoil of some forest stands in central Sweden. *Canadian Journal of Forestry Research*, **15**, 715–22.

Notohadiprawiro, T. (1981). Peat deposition, an idle stage in the natural cycling of nitrogen, and its possible activation for agriculture. In *Nitrogen Cycling in S.E. Asian Wet Monsoonal Ecosystems*, ed. R. Wetselaar, J. R. Simpson & T. Rosswall, pp. 139–47. Canberra: Australian Academy of Science.

O'Brien, W. J. (1981). Use of aquatic macrophytes for wastewater treatment. *Proceedings of the American Society for Civil Engineering*, **107**, 681–98.

Okon, Y. & Kapulnik, Y. (1985). Development and function of *Azospirillum*-inoculated roots. *Plant and Soil*, **90**, 3–16.

Olson, R-J. (1980). *Studies of Biological Nitrogen Cycle Processes in the Upper Waters of the Ocean with Special Reference to the Primary Nitrite Maximum*. Ph.D. Thesis, University of California, San Diego. (quoted by Kaplan, 1983).

Pate, J. S. & Dixon, K. W. (1981). Plants with fleshy underground storage organs – a

Western Australian survey. In *The Biology of Australian Plants*, ed. J. S. Pate and A. J. McComb, pp. 181–215. Nedlands: University of Western Australia Presss.

Patriquin, D. G. (1986). Biological husbandry and the 'nitrogen problem'. *Biological Agriculture and Horticulture*, **3**,, 167–89.

Patriquin, D. G. & McClung, C. R. (1978). Nitrogen accretion, and the nature and possible significance of N$_2$ fixation (acetylene reduction) in Nova Scotian *Spartina alterniflora* stands *Marine Biology*, **47**, 227–42.

Payne, W. J. (1981). *Denitrification*. New York: Wiley-Interscience.

Pfanz, H. & Heber, U. (1986). Buffer capacities of leaves, leaf cells and leaf cell organelles in relation to fluxes of potentially acidic gases. *Plant Physiology*, **81**, 597–602.

Plazinski, J. & Rolfe, B. G. (1985a). Interaction of *Azospirillum* and *Rhizobium* strains leading to inhibition of nodulation. *Applied and Environmental Microbiology*, **49**, 990–3.

Plazinski, J. & Rolfe, B. G. (1985b). Influence of *Azospirillum* strains on the nodulation of clovers by *Rhizobium* strains. *Applied and Environmental Microbiology*, **49**, 984–9.

Polhill, R. M. & Raven, P. H. (ed.) (1981). *Advances in Legume Systematics, Part 1*. Kew: Rotal Botanic Gardens.

Ponder, F. & Tadros, S. H. (1985). Juglone concentration in soil beneath black walnut interplanted with nitrogen-fixing species. *Journal of Chemical Ecology*, **11**, 937–42.

Postgate, J. R. (1982). *The Fundamentals of Nitrogen Fixation*. Cambridge: Cambridge University Press.

Postgate, J. R. (1987). *Nitrogen fixation*, 2nd edn Studies in Biology, no. 92. London: Edward Arnold.

Prestwick, G. D. & Bentley, B. L. (1981). Nitrogen fixation by intact colonies of the termite *Nasutitermes cornegia. Oecologia*, **49**, 249–51.

Price, N. M., Cochlan, W. P. & Harrison, P. J. (1985). Time course of uptake of inorganic and organic nitrogen by phytoplankton in the Strait of Georgia: comparison of frontal and stratified communities. *Marine Ecology – Progress Series*, **27**, 39–53.

Probert, M. F. & Williams, J. (1985). The residual effectiveness of phosphorus for *Stylosanthes* pastures in red and yellow earths in the semi-arid tropics. *Australian Journal of Soil Research*, **23**, 211–22.

Prosser. J. I. (ed.) (1986). *Nitrification*, Special Publications of the Society for General Microbiology, vol. 20. Oxford: IRL Press.

Raison, R. J. (1979). Modification of the soil environment by vegetation fires, with particular reference to nitrogen transformations: a review. *Plant and Soil*, **51**, 73–108.

Rao, D. L. N. & Ghai, S. K. (1986). Urease inhibitors: effect on wheat growth in an alkali soil. *Soil Biology and Biochemistry*, **18**, 255–8.

Raven, J. A. (1977). The evolution of vascular land plants in relation to supercellular transport processes. *Advances in Botanical Research*, **5**, 153–219.

Raven, J. A. (1985). Regulation of pH and generation of osmolarity in vascular plants: a cost benefit analysis in relation to efficiency of use of energy, nitrogen and water. *New Phytologist*, **101**, 25–77.

Raven, J. A. & Smith, F. A. (1976). Nitrogen assimilation and transport in vascular plants in relation to intracellular pH regulation. *New Phytologist*, **76**, 205–12.

Reddy, K. R. & Patrick, W. H. (1986). Denitrification losses in flooded rice fields. *Fertilizer Research*, **9**, 99–116.

Reiners, W. A. (1983). Transport processes in the biogeochemical cycles of carbon, nitrogen, phosphorus and sulphur. In *The Major Biogeochemical Cycles and their Interactions*. Scope 21, ed. B. Bolin & R. B. Cook, pp. 115–41. Chichester: John Wiley & Sons.

Reuter, D. J. & Robinson, J. B. (eds.) (1986). *Plant Analysis – An Interpretation Manual*. Melbourne: Inkata Press.

Rho, J. (1986). Microbial interactions in heterotrophic nitrification. *Canadian Journal of Microbiology*, **32**, 243–7.

Rice, E. L. & Pancholy, S. K. (1974). Inhibition of nitrification by climax ecosystems. III. Inhibitors other than tannins. *American Journal of Botany*, **61**, 1095–1103.

Richardson, C. J., Tilton, D. L., Kadlec, J. A., Chamie, J. P. M. & Wentz, W. A. (1978). Nutrient dynamics of northern wetland ecosystems. In *Freshwater Wetlands*, ed. R. E. Good, D. F. Whigham & R. L. Simpson, pp. 217–41. New York: Academic Press.

Richey, J. E. (1983). The phosphorus cycle. In *The Major Biogeochemical Cycles and their Interactions*. Scope 21, ed. B. Bolin & R. B. Cook, pp. 51–6. Chichester: John Wiley & Sons.

Robinson, D. & Rorison, I. H. (1983). Relationships between root morphology and nitrogen availability in a recent theoretical model describing nitrogen uptake in soils. *Plant, Cell and Environment*, **6**, 641–7.

Robson, A. D. (1983). Mineral nutrition. In *Nitrogen Fixation, Volume 3: Legumes*, ed. W. J. Broughton, pp. 36–55. Oxford: Clarendon Press.

Rosswall, T. (1981). The biogeochemical nitrogen cycle. In *Some Perspectives of the Major Biogeochemical Cycles*, Scope 17, ed. G. E. Likens, pp. 25–49. Chichester: John Wiley & Sons.

Rosswall, T. (1982). Microbiological regulation of the biochemical nitrogen cycle. *Plant and Soil*, **67**, 15–34.

Rosswall, T. (1983). The nitrogen cycle. In *The Major Biogeochemical Cycles and their Interactions*. Scope 21, ed. B. Bolin & R. B. Cook, pp. 46–50. Chichester: John Wiley & Sons.

Rosswall, T. & Granhall, U. (1980). Nitrogen cycling in a subarctic ombrotrophic mire. In *Ecology of a Subarctic Mire*, Ecological Bulletin 30, ed. M. Sonesson, pp. 209–34. Stockholm: Swedish National Science Research Council.

Rubio, J. L. & Hauck, R. D. (1986). Uptake and use patterns of nitrogen from urea, oxamide, and isobutylidene diurea by rice plants. *Plant and Soil*, **94**, 109–23.

Rueter, J. G. (1982). Theoretical Fe limitations of microbial N_2 fixation in the oceans. *EOS*, **63**, 945.

Salati, E., Sylvester-Bradley, R. & Victoria, R. L. (1982). Regional gains and losses of nitrogen in the Amazonian basin. *Plant and Soil*, **67**, 367–76.

Scaglia, J., Lensi, R. & Chalamet, A. (1985). Relationship between photosynthesis and denitrification in planted soil. *Plant and Soil*, **84**, 37–43.

Scranton, M. I. (1983). Gaseous nitrogen compounds in the marine environment. In *Nitrogen in the Marine Environment*, ed. E. J. Carpenter & D. G. Capone, pp. 37–64. New York: Academic Press.

Seastedt, T. R. (1985). Maximization of primary and secondary productivity by grazers. *The American Naturalist*, **126**, 559–64.

Seneviratne, R. & Wild, A., (1985). Effect of mild drying on the mineralization of soil nitrogen. *Plant and Soil*, **84**, 175–9.

Sharma, B. & Ahler, R. C. (1977). Nitrification and nitrogen removal. *Water Research*, **11**, 897–925.

Sharp, J. H. (1983). The distribution of inorganic nitrogen and dissolved and particulate organic nitrogen in the sea. In *Nitrogen in the Marine Environment*, ed. E. J. Carpenter & D. G. Capone, pp. 1–35. New York: Academic Press.

Shipton, W. A. & Burggraaf, A. J. P. (1982). *Frankia* growth and activity as influenced by water potential. *Plant and Soil*, **69**, 293–7.

Sinclair, T. R. & Goudriaan, J. (1981). Physiological and morphological constraints on transport in nodules. *Plant Physiology*, **67**, 143–5.

Skiba, U. & Wainwright, M. (1984). Nitrogen transformation in coastal and dune soils. *Journal of Arid Environments*, **7**, 1–8.

Skinner, F. A. & Uomala, P. (eds.) (1986). *Nitrogen Fixation with Non-Legumes*. Dordrecht: Martinus Nijhoff.

Skujiņš, J. (1981). Nitrogen cycling in arid ecosystems. In: *Terrestrial Nitrogen Cycles*, Ecological Bulletin 33, ed. F. E. Clark & T. Rosswall, pp. 477–91. Stockholm: Swedish Natural Science Research Council.

Skujiņš, J. (1984). Microbial ecology of desert soils. *Advances in Microbial Ecology*, 7, 49–91.

Smith, M. S. (1982a). Dissimilatory reduction of NO_2^- to NH_4^+ and N_2O by a soil *Citrobacter* sp. *Applied and Environmental Microbiology*, 43, 854–60.

Smith, O. L. (1982b). *Soil Microbiology: A Model of Decomposition and Nutrient Cycling*. Boca Raton: CRC Press Inc.

Smith, S. J. & Young, L. B. (1975). Distribution of nitrogen forms in virgin and cultivated soils. *Soil Science*, 120, 354–60.

Spencer, C. P. (1975). The micronutrient elements. In *Chemical Oceanography*, vol. 2, 2nd edn, ed. J. P. Riley & G. Skirrow, pp. 245–399. London: Academic Press.

Sprent, J. I. (1979). *The Biology of Nitrogen Fixing Organisms*. Maidenhead: McGraw-Hill.

Sprent, J. I. (1985). Nitrogen fixation in arid environments. In *Plants for Arid Lands*, ed. G. E. Wickens, J. R. Goodin & D. V. Field, pp. 215–29. London: George Allen & Unwin.

Sprent, J. I. (1986a). Potential for nitrogen fixing legume trees in the tropics. *International Tree Crops Journal*, 4, 47–54.

Sprent, J. I. (1986b). Benefits of *Rhizobium* to agriculture. *Trends in Biotechnology*, 4, 124–9.

Sprent, J. I. (1987). Problems and potentials for nitrogen fixation in deserts. In *Arid Lands: Today and Tomorrow*, ed. G. P. Nabham & A. Elias-Cesnik. Tucson: University of Arizona Press, in press.

Sprent, J. I. & Minchin, F. R. (1985). *Rhizobium* nodulation and nitrogen fixation. In *Grain Legume Crops*, ed. E. R. Roberts & R. J. Summerfield, pp. 115–44. London: Granada Technical Books.

Sprent, J. I. & Raven, J. A. (1985). Evolution of nitrogen-fixing symbioses. *Proceedings of the Royal Society of Edinburgh*, 85(B), 215–37.

Sprent, J. I., Sutherland, J. M. & Faria, S. M. de (1987a). Some aspects of the biology of nitrogen fixing organisms. *Philosophical Transactions of the Royal Society London*, B317, in press.

Sprent, J. I., Sutherland, J. M. & Faria, S. M. de (1987b). Structure and function of root nodules from woody legumes. In *The Biology of Legumes*, ed. C. H. Stirton & J. L. Zarucchi. St Louis: Missouri Botanical Gardens, in press.

Stackebrandt, E. & Woese, C. R. (1984). The phylogeny of prokaryotes. *Microbiological Science*, 1, 117–22.

Stanford, G., Legg, J. O., Dzienia, S. & Simpson, E. C. (1975). Denitrification and associated nitrogen transformations in soils. *Soil Science*, 120, 147–52.

Stanford, C. R., Vander Pol, R. A. & Dzienia, S. (1975). Denitrification rates in relation to total and extractable soil carbon. *Soil Science Society of America Proceedings*, 39, 284–9.

Stewart, W. D. P. & Alexander, G. (1971). Phosphorus availability and nitrogenase activity in aquatic blue–green algae. *Freshwater Biology*, 1, 389–404.

Stewart, W. D. P., Preston, T., Peterson, H. G. & Christofi, N. (1982). Nitrogen cycling in eutrophic freshwaters. *Philosophical Transactions of the Royal Society London*, B, 296, 491–509.

Stewart, W. D. P. & Rowell, P. (1986). Biochemistry and physiology of nitrogen fixation with particular emphasis on nitrogen-fixing phototrophs. *Plant and Soil*, 90, 167–91.

Stock, W. D. & Lewis, O. A. M. (1986). Soil nitrogen and the role of fibre as a mineralizing agent in a South African coastal fynbos ecosystem. *Journal of Ecology*, 74, 317–28.

Stojkovski, S., Payne, R., Magee, R. J. & Stanisich, V. A. (1986). Binding of molybdenum to slime produced by *Pseudomonas aeruginosa* PAO1. *Soil Biology and Biochemistry*, **18**, 117–18.

Stowell, R., Ludwig, R., Colt, J. & Tchobanoglous, G. (1981). Concepts in aquatic treatment system design. *Journal of Environmental Engineering Division, Proceedings of American Society of Civil Engineering*, **107**, 919–40.

Tam,T-Y. & Knowles, R. (1979). Effects of sulphide and acetylene on nitrous oxide reduction by soil and by *Pseudomonas aeruginosa*. *Canadian Journal of Microbiology*, **25**, 1133–8.

Tilak, K. V. B. R., Singh, C. S. & Rana, J. P. S. (1981). Effects of combined inoculation of *Azospirillum brasilense* with *Rhizobium trifolii, Rhizobium meliloti* and *Rhizobium* sp. (cowpea miscellany) on nodulation, and yield of clover (*Trifolium repens*), lucerne (*Medicago sativa* and chick pea (*Cicer arietinum*). Zentralblatt für Bakteriologie, Parasitenkunde. Infektionskrankheiten und Hygiene, Part II, **136**, 117–20.

Towe, K. M. (1985). Habitability of the early earth: clues from the physiology of nitrogen fixation and photosynthesis. *Origins of Life*, **15**, 235–50.

Triska, F. J., Sedell, J. R., Cromack, K., Gregory, S. V. & McCorison, F. M. (1984). Nitrogen budget for a small coniferous forest stream. *Ecological Monographs*, **54**, 119–40.

Upadhyay, V. P. & Singh, J. S. (1985). Nitrogen dynamics of decomposing hardwood leaf litter in a central Himalayan forest. *Soil Biology and Biochemistry*, **17**, 827–30.

Veal, D. A. & Lynch, J. M. (1984). Associative cellulolysis and dinitrogen fixation by co-cultures of *Trichoderma harzianum* and *Clostridium butyricum*. *Nature*, **310**, 695–7.

Vincent, W. F., Downes, M. T. & Vincent, C. L. (1981). Nitrous oxide cycling in Lake Vanda, Antarctica. *Nature*, **292**, 618–20.

Vitousek, P. M. (1984). Litterfall, nutrient cycling, and nutrient limitation in tropical forests. *Ecology*, **65**, 285–98.

Vitousek, P. M., Gosz, J. R., Greir, C. L., Melillo, J. M. & Reiners, W. A. (1982). A comparative analysis of potential nitrification and nitrate mobility in forest ecosystems. *Ecological Monographs*, **52**, 155–77.

Vogt, K. A., Grier, C. C. & Vogt, D. J. (1986). Production, turnover, and nutrient dynamics of above- and belowground detritus of wood forests. In *Advances in Ecological Research*, ed. A. MacFayden & E. D. Ford, pp. 303–78. London: Academic Press; Harcourt Brace Jovanovich.

Wallace, W. (1986). Distribution of nitrate assimilation between the root and shoot of legumes and a comparison with wheat. *Physiologia Plantarum*, **66**, 630–6.

Waterbury, J. B., Calloway, C. D. & Turner, R. D. (1983). A cellulolytic nitrogen-fixing bacterium from the gland of Deshayes in shipworms (Bivalvia: teredinidae). *Science*, **221**, 1401–3.

Watson, S. W., Valois, F. W. & Waterbury, J. B. (1981). The family Nitrobacteriaceae. In *The Prokaryotes*, ed. M. P. Starr, H. Stolp, H. G. Trüpes, A. Balows & H. G. Schlegel, pp. 1005–22. Berlin: Springer-Verlag.

Weathers, P. J. (1984). N₂O evolution by green algae. *Applied and Environmental Microbiology*, **48**, 1251–3.

Weathers, P. J. & Niedzielski, J. J. (1986). Nitrous oxide production by cyanobacteria. *Archives of Microbiology*, **146**, 204–6.

Webb, K. L. & Wiebe, W. J. (1975). Nitrification on a coral reef. *Canadian Journal of Microbiology*, **21**, 1427–31.

West, N. E. & Skujiņš, J. (1977). The nitrogen cycle in North American cold-winter and semi-desert ecosystems. *Oecologia Plantarum*, **12**, 45–53.

Wetselaar, R. (1980). Nitrogen cycling in a semi-arid region of tropical Australia. In *Nitrogen Cycling in West African Ecosystems*, ed. T. Rosswall, pp. 157–69. Uppsala: SCOPE/UNEP.

Wharton, R. A., McKay, C. P., Mancinelli, R. L. & Simmons, G. M. (1987). Perennial N_2 supersaturation in an Antarctic lake. *Nature*, **325**, 343–5.

Whelan, A. M. & Alexander, M. (1986). Effect of low pH and high Al, Mn and Fe levels on the survival of *Rhizobium trifolii* and the nodulation of subterranean clover. *Plant and Soil*, **92**, 363–79.

Whiting, G. J., Gandy, E. L. & Yoch, D. C. (1986). Tight coupling of root-associated nitrogen fixation and plant photosynthesis in the salt marsh grass *Spartina alterniflora* and carbon dioxide enhancement of nitrogenase activity. *Applied and Environmental Microbiology*, **52**, 108–13.

Wickens, G. E., Goodin, J. R. & Field, D. V. (1985). *Plants for Arid Lands*. London: George Allen & Unwin.

Wilkinson, W. B. & Greene, L. A. (1982). The water industry and the nitrogen cycle. *Philosophical Transactions of the Royal Society London, B*, **296**, 459–75.

Williams, B. L., Cooper, J. M. & Pyatt, D. G. (1979). Some effects of afforestation with Lodgepole Pine on rates of nitrogen mineralization in peat. *Forestry*, **52**, 151–60.

Williams, P. M. (1967). Sea surface chemistry: organic and inorganic nitrogen and phosphorus in surface films and surface waters. *Deep Sea Research*, **14**, 791–800.

Wolverton, B. C., McDonald, R. C. & Duffer, W. R. (1983). Microorganisms and higher plants for waste water treatment. *Journal of Environmental Quality*, **12**, 236–42.

Wong-Chong, G. M. & Loehr, R. C. (1975). The kinetics of microbial nitrification. *Water Resources*, **9**, 1099–1106.

Wood, P. M. (1986). Nitrification as a bacterial energy source. In *Nitrification*, Special Publications of the Society for General Microbiology, vol. 20, ed. J. I. Prosser. pp. 39–62. Oxford: IRL Press.

Yordy, D. M. & Ruoff, K. L. (1981). Dissimilatory nitrate reduction to ammonia. In *Denitrification, Nitrification and Atmospheric Nitrous Oxide*, ed. C. C. Delwiche, pp. 171–91. New York: John Wiley & Sons, Inc.

Young, C. P., Oakes, D. B. & Wilkinson, W. B. (1979). The impact of agricultural practices on the nitrate content of groundwater in the principal United Kingdom aquifers. In *International Institute for Applied Systems Analysis International Conference on Environmental Management of Agricultural Watersheds*, Smolenice, Czechoslovakia, April 1979.

Zahran, H. H. & Sprent. J. I. (1986). Effects of sodium chloride and polyethylene glycol on root-hair infection and nodulation of *Vicia faba* L. plants by *Rhizobium leguminosarum*. *Planta*, **167**, 303–9.

Index

Only the major genera and species cited in the text are indexed.